T0269063

NEUROSCIENCE BASICS

NEUROSCIENCE BASICS

A Guide to the Brain's Involvement in Everyday Activities

JENNIFER L. LARIMORE

Agnes Scott College, Decatur, Georgia, United States

ACADEMIC PRESS

An imprint of Elsevier

Academic Press is an imprint of Elsevier
125 London Wall, London EC2Y 5AS, United Kingdom
525 B Street, Suite 1800, San Diego, CA 92101-4495, United States
50 Hampshire Street, 5th Floor, Cambridge, MA 02139, United States
The Boulevard, Langford Lane, Kidlington, Oxford OX5 1GB, United Kingdom

Notices

Knowledge and best practice in this field are constantly changing. As new research and experience broaden our understanding, changes in research methods, professional practices, or medical treatment may become necessary.

Practitioners and researchers must always rely on their own experience and knowledge in evaluating and using any information, methods, compounds, or experiments described herein. In using such information or methods they should be mindful of their own safety and the safety of others, including parties for whom they have a professional responsibility.

To the fullest extent of the law, neither the Publisher nor the authors, contributors, or editors, assume any liability for any injury and/or damage to persons or property as a matter of products liability, negligence or otherwise, or from any use or operation of any methods, products, instructions, or ideas contained in the material herein.

British Library Cataloguing-in-Publication Data
A catalogue record for this book is available from the British Library

Library of Congress Cataloging-in-Publication Data
A catalog record for this book is available from the Library of Congress

ISBN: 978-0-12-811016-4

For Information on all Academic Press publications
visit our website at https://www.elsevier.com/books-and-journals

Working together
to grow libraries in
developing countries

www.elsevier.com • www.bookaid.org

Publisher: Nikki Levy
Acquisition Editor: Natalie Farra
Editorial Project Manager: Kathy Padilla
Production Project Manager: Laura Jackson
Designer: Victoria Pearson Esser

Typeset by MPS Limited, Chennai, India

DEDICATION

This book is dedicated to my students who challenge me to be a better professor, to the lab mates and colleagues who have challenged me to be a better neuroscientist, my family who has supported and enhanced my education, my daughters who remind me of what really matters in life, and to my husband for the endless support and love through my career. Thank you all!

CONTENTS

PREFACE

Maybe you are one of those nerds who loves listening to podcasts about the latest scientific discoveries and you want to better understand how the brain works. Or maybe you are walking alongside a loved one diagnosed with brain disorder that you would like to understand a little better. Maybe you are one of those science nerds trying to get a head start on your introduction to neuroscience class.

No matter what is driving your interest in the brain, you've come to a good space to discover more about the brain outside of the classroom. There are many amazing textbooks that guide students through the basics of neurobiology. But this book is aimed at guiding anyone interested in how the brain works through some of the basic topics covered in an introductory neuroscience class—minus a lot of the jargon, quizzes, tests, projects, and mathematical equations.

Another way to view this books is a "cliff notes" version of the major textbooks used to teach introductory neuroscience. If you ever want to dive deeper, there are references at the end of each chapter that can help you with that.

Our days are intricate, and activities during the day are also complex. So how does our brain work throughout a typical day? How does the brain function during some of these basic aspects that we encounter in a daily routine?

We will discuss what normal function looks like and what happens when things don't go according to plan. When we explore some of the diseases, I will mention some movies or documentaries that chronicle that disease. While reading about the disease symptoms and what occurs in the brains of people with these diseases helps our understanding, these documentaries and movies allow us to see the human side of the disease and what it is like to a person with that disorder. And the movies listed are in no way an exhaustive list.

On the science end of this book, I will unapologetically use the metric system in this book. I live in the metric system. I weigh things in grams and my glassware in my lab is in the metric system. Scientists love acronyms. The hard part with science acronyms is figuring out when you pronounce it like a word, or individually say each letter. I will point that out throughout this book.

And who am I that I can offer light on these topics? Good question! I always encourage my students to consider the source of the information, and you should do nothing less. I am a scientist and a professor at a liberal arts college in Atlanta, Georgia. My first time touching a brain was in high school. I was so nervous that my palms were a little sweaty, which made the powder in the latex gloves create a paste on my skin. There in my hands was a human brain. It was so amazing, squishy yet not falling apart, and so complex. While the smell in the cadaver lab reeked of formaldehyde which made me a little light headed, I was hooked on studying the brain. *(Shout out to my AP Biology teacher, Mrs. Bobbitt! Thank you for introducing me to such a fascinating area of study!)*

When I was in high school and college, I thought I wanted to be a physician. As I shadowed physicians, I realized that my brain did not really think the way physicians think. I started working for a science lab after college until I could figure out what I wanted to do. In that lab, I realized how amazing scientific research was and that my brain was definitely wired to think like a researcher. I applied to a Ph.D. program and after 5 years of training. I graduated with my Ph.D. in Neurobiology. I did postdoctoral training in the cell biology of neurodevelopmental disorders as well as in science education. After my post-doctoral fellowship, I began teaching and conducting research at the small liberal arts college I am at today. All that to say, I have sat under great instructors who have explained these topics well and I have taught these topics at the undergraduate level.

My love of the brain is not first generation. I am at least a second-generation brain lover (in a non-zombie way, of course). My dad researches clinical anesthesia. My mom studies and teaches the psychology of early childhood education. And, the cherry on top of my neuroscience lineage is a pet. We had a family pet named after a structure of the of brain, the synapse. I have a rich nerd lineage.

Now, let's learn about what the brain does everyday.

CHAPTER 1

How to Build a Human Brain

Contents

SUMMARY

In order for us to understand how the brain works during a normal day, we must first understand what is a brain and how does the brain develop during pregnancy and childhood. During brain development, the brain is organized so that the daily functions carried out by the brain are organized and efficient. Finally, we will cover a few of the disorders that can occur when development does not go according to plan.

Neuroscience Basics.
DOI: http://dx.doi.org/10.1016/B978-0-12-811016-4.00001-5

1.1 WHAT IS THE BRAIN?

In order to understand how the brain works, we need to understand what the brain is. The term brain usually refers to the tissue found within a skull that generates behavior. The mind usually refers to personality, opinions, experiences, and memories. There is an age-old philosophical debate arguing the brain and the mind are the same thing versus they are separate things. Aristotle (384−322 BCE) and Plato (428−348 BCE) argued that the soul of a person was the center of intelligence or wisdom and that the soul was not something physical; therefore, the mind and the brain were different. Descartes (1596−1650) was a French philosopher who believed that the mind controlled consciousness and self-awareness and was separate from the brain, but the brain was the seat of intelligence. Another side of the debate is given by a researcher in the field of artificial intelligence (AI), Marvin Minsky (1927−2016). In the early 1970s, Minsky was working in an AI lab at MIT where he developed with others the Society of the Mind Theory. The theory was published in Minsky's book in 1986—*The Society of the Mind*. Minsky stated that "minds are simply what the brains do." There are many more philosophers and researchers who have weighed in on this debate spanning several centuries. While this interesting debate continues, we will direct our focus on the physical tissue inside the skull, the brain.

In order to understand how the brain functions properly through the day, we need to first understand how the brain develops during pregnancy and childhood.

1.2 OVERVIEW OF BRAIN DEVELOPMENT

Have you ever tried to follow a professional's recipe? Trying to recreate something a professional dreamed up is not an easy task. In a similar fashion, brain development must follow its recipe; otherwise, problems can arise. There are many steps to making a brain, and each of them is tightly regulated. The remainder of this chapter will cover the five steps to making a brain; how childhood is necessary for brain development; how the brain organizes itself during development; and what happens when brain development doesn't go according to plan.

The majority of the brain's "thinking cells," the neurons, are already in place in the womb. In the womb, the brain weighs less than 300 grams.

Directly after birth, the brain triples in size and other cells the brain needs to do its task have been put into place. By adulthood, the human brain weighs between 1300−1400 grams.

There are five main steps in growing a brain:

Creation and closing of the neural tube (neurulation)

Birth and growth of precursor cells

Cells find their proper location (migration)

Precursor cells figure out what kind of cell they are going to be (differentiation)

Connections between different cells are made—the brain is starting to be wired together (targeting/synaptogenesis).

1.3 BRAIN DEVELOPMENT STEPS

1.3.1 Step 1: Creation and Closing of the Neural Tube (Neurulation)

Humans all start off as just two cells—an egg and a sperm. Eventually, those cells divide and create four cells, then eight, doubling constantly. Eventually, the cells make something of a sheet, initially with two layers. This two-layer sheet (bilaminate epiblast) occurs during day 14−15 gestation in humans. This two-layer sheet becomes a three-layer sheet in a process called gastrulation. The three layers are: endoderm, ectoderm, and mesoderm. The endoderm becomes the epidermis and associated structures as well as the nervous system. The ectoderm becomes the glands and the inner lining for both the digestive and respiratory system. Finally, the mesoderm becomes the muscles, bones, cartilage, circulatory system, excretory system, gonads, and the outer covering of the internal organs. As this three-layer ball starts to organize, it will form a tube that closes. The open tube is formed around day 18−19 gestation in humans and begins the folding around day 21 and fusing to form a closed tube around day 27 gestation. When the neural tube fails to close properly, diseases like spina bifida or brain hernias can result. About 2 days after full closure of the neural tube, arm buds begin to form. This entire process, just 27 days of the gestation, are intense and tightly regulated in order to regulate proper neural tube closure.

At day 21, as the neural tube is closing, the tube begins specialization. The tube forms three regions: the prosencephalon, the mesencephalon, and the rhombencephalon (Fig. 1.1). The prosencephalon becomes the forebrain, the mesencephalon becomes the midbrain, and the

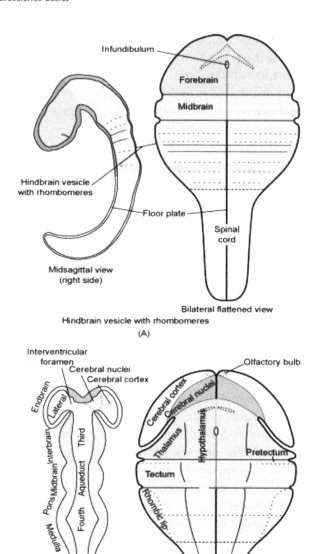

Figure 1.1 (A) the prosencephalon (forebrain), the mesencephalon (midbrain), and the rhombencephalon (hindbrain). (B) Telecephalon (cortex, hippocampus, basal ganglia, and olfactory bulb), diencephalon (thalamus, hypothalamus, and optic cups), mesencephalon (superior and inferior colliculi, and midbrain tegmenjtum), metencephalon (cerebellum and pons), and the myelencephalon (medulla).

rhombencephalon becomes the hindbrain. So 3—4 weeks into gestation, the brain is already well on its way! Precursors are in place and filled with the cells necessary for further development.

At this point in gestation, most humans are unaware or at best, are unsure if they are pregnant. Most women begin to question if they are pregnant due to a missed menstrual cycle. The menstrual cycle should usually occurs 1—2 weeks (7—14 days) after gestation. There is a tighter range that is more "textbook" but some women have ovulation cycles that are not textbook. In order for pregnancy to continue, hormone levels must rise. These elevated hormone levels are utilized in pregnancy tests and detected in the urine. Currently, some home pregnancy tests boast that they will detect a pregnancy 5 days before a period. Most tests are less reliable the earlier they are used. Usually, a test is reliable 1 week after a missed menstrual cycle, which would put a woman at 2—3 weeks (14—21 days) gestation around the time of the test.

By 8 weeks of gestation, the prosencephalon has further organized into the telencephalon and diencephalon. The mesencephalon is still present and the rhombencephalon becomes the metencephalon and myelencephalon (Fig. 1.1). Each of those "cephalons" later become a region of the brain. The telencephalon later becomes the cortex, hippocampus, basal ganglia, and olfactory bulb, the diencephalon becomes the thalamus and the hypothalamus as well as the optic cups, the mesencephalon becomes the midbrain, the metencephalon becomes the pons and the cerebellum and the myelencephlon becomes the medulla (Fig. 1.1). The bottom line is, at 8 weeks, the precursors to all the major brain regions are in place.

1.3.2 Step 2: Birth and Growth of Cells That Could Become a Lot of Things (Precursor Cells) and Step 3: Cells Find Their Proper Location (Migration)

There are precursor cells, called stem cells, which generate the types of cells found in the brain. Once all the cells are born and they move to the region where they can carry out their specific function and the whole nervous system gets organized based on what functions are to be carried out, a brain has been formed. These migrations are aided by other cells (namely, radial glia) and signals that attract or repel new cells. These steps occur from about the fifth week of gestation to 7 months postnatally.

1.3.3 Step 4: Precursor Cells Figure Out What Kind of Cell They Are Going to Be (Differentiation)

At this point (usually at 25 weeks gestation to 5 months postnatally), the brain is creating more cells in order for the brain to function properly. It needs a large army to keep you working every day. More stem cells are called in and they respond to different cues in order to figure out what their function is in a brain. There are signals inside and outside the cell which tell the cell who it needs to be in order for the brain to function properly.

1.3.4 Step 5: Connections Between Different Cells Are Made—The Brain Is Starting to Be Wired Together (Targeting/Synaptogenesis)

Most of brain connections (synapses) are solidified during the formative years of childhood. Experiences and learning impact the developing brains in ways scientists are still exploring. Eventually, the brain stops growing and dials back (pruning) the growth just a touch to make sure that proper and strong connections are established between brain regions and communication between the brain's thinking cells (neurons) is correct. The brain prunes back some of the connections so that the brain is not "overwired." Too many connections in the brain are the equivalent of too many chefs in the kitchen. The proper number and strength of connections are important in development. There cannot be too many or too few. Most of the pruning and strengthening of the connections occur between 8 and 12 years of age. The last part of the brain to develop is a region that is necessary for assessing risk, personality, and long-term memory—the prefrontal cortex, which matures in the mid-20s, when driver insurance rates drop and you can rent a car.

The ability to change connections, called plasticity, is a characteristic of the human brain throughout a lifetime. The connections in the brain that are made are called synapses. So the ability of the brain to change is called synaptic plasticity. Because the brain is plastic, humans are able to learn and are also able to forget.

We have defined what a brain is and we have discussed the basic steps of human brain development. In this section, we have learned that connections are being formed and changed throughout childhood. We have also learned that many of the stem cells are determining their fate after birth. So what role does childhood serve in brain development? How does our childhood experiences impact our everyday adult life?

1.4 WHAT PURPOSE DOES HUMAN CHILDHOOD SERVE? BACK TO THE SANDBOX

Connections (synapses) between the brain's thinking cells (neurons) are being made through late childhood/prepuberty. What is fascinating here is that humans have this time of brain growth *after* birth. And this time of brain growth after birth allows interactions with the environment and experiences with other humans to begin shaping personalities. During the evolutionary process, humans have gained approximately 6 years of childhood, a developmental period that the majority of other hominids do not have. Adding that time of childhood lengthens the time between birth and mating, which slows down a species ability to continue, which is why other hominids do not have a childhood.

The evolved, larger brains found in current humans resulted in neoteny (from the Greek—*neos* means "new or juvenile" and *teinein* means to "extend"), which refers to the retention of youthful traits. Youthful traits here refer to childhood, not a search for the fountain of youth or botox on a wrinkled brow. Why do humans have this extended youth? Think about what neoteny might mean for child birth. Bigger brains mean bigger skulls. With a growing brain, pregnancy has two options—to have a longer gestation and give birth to a much larger baby with a larger skull (an option no pregnant lady will ever choose), or to continue with the current human gestation and give birth to a more helpless and less-developed baby than primate cousins carried to term. The human species opted for the shorter gestation and less formed baby, much to the relief of anyone who has given birth.

Because of this, the same brain networks that are fully developed at birth for other hominids are still being wired for humans *after* birth based on experiences, learning, and memory. In childhood, humans are developing personality and the ability to generate behavior. During those years, the brain is wiring and rewiring, allowing the brain to become better suited to the environment. During this period of wiring and rewiring and during the later stages of development in the womb (migration and differentiation), the brain organizes itself. Before the work of Pierre Paul Broca (1824—80) in the mid-1800s, it was thought the brain did not localize functions. Meaning, there was no one part of the brain responsible for functions like language or memory. Work by Broca and others demonstrated that indeed regions of the brain were necessary for particularly functions. Based on their work, scientists were better able to understand the organization of the nervous system and how that organization occurred during development.

1.5 ORGANIZATION OF THE NERVOUS SYSTEM

During development, when cells are migrating and differentiating, they are organizing regions that are responsible for particular functions. Because of these distinct regions, we can better understand the overall organization of the brain. While the organization of the nervous system can be a bit dry, it is necessary to understand the regions and their roles in function because it gives us a better understanding of what symptoms to expect if that region is damaged due to stroke, brain injury, or loss of connections.

The brain has several key components and pieces. The brain belongs to the nervous system. There are two main divisions to our nervous system: the peripheral (noncentral) nervous system (PNS—each letter said individually) and the central nervous system (CNS—each letter said individually). The PNS has two mains parts. One is the sensory thinking cells (sensory neurons) that detect touch or pressure or heat on the surface of the body or within our body. The other main division of the PNS is motor thinking cells (motor neurons) that regulate things like planned movements (walking) or unplanned movements (flight or fight).

The brain belongs to the CNS (as does the spinal cord). The CNS is responsible for analyzing and integrating all the information from the sensory and motor information collected by the PNS.

1.5.1 The Seven Major Divisions of the CNS

There are seven major divisions of the CNS, each with its own unique set of functions. Each of these seven divisions are established by migration and differentiation that occur during brain development. Without proper cell types or cell migration, these areas would not function properly during childhood or adulthood. These are also the major divisions that will determine normal everyday activities that we will highlight throughout the book.

1. The spinal cord extends from the base of the skull to the first lumbar vertebra. It takes in sensory information from the body and it contains motor thinking cells (neurons) that are responsible for voluntary and reflexive movement.

2. The medulla regulates blood pressure, respiration, taste, hearing, balance, and controlling facial muscles.
3. The pons is necessary for movement, sensation, respiration, taste, and sleep.
4. The midbrain is an important relay station for motor control and regulates several components of the auditory and visual systems.
5. The cerebellum is necessary for movement, maintaining posture, coordinating head-and-eye movements, fine movements and is implicated in language and cognition.
6. The diencephalon contains the thalamus and the hypothalamus. The thalamus determines what sensory information reaches awareness, integrates motor information, and influxes levels of attention. The hypothalamus regulates eating, growth, drinking, and maternal behaviors. It is an essential part of the motivation system of the brain.
7. Finally, the cerebral hemispheres. The brain has two hemispheres (right and left). These two sides are joined in the middle (a region called the corpus callosum). The left hemisphere regulates logical functions and is responsible for words, numbers, analysis, lists, linearity, and sequence. The left hemisphere also controls the right side of the body. The right hemisphere is the more creative. It is responsible for rhythm, spatial awareness, imagination, daydreaming, and holistic awareness. The right side of the brain controls the left side of the body.

1.5.2 The Four Lobes

The seventh division of the CNS, the cerebral hemispheres, can be further divided into the 4 lobes. These four lobes are distinguishable during development and organized during migration and differentiation highlighting the need for a highly regulated developmental process for the human brain (Fig. 1.2).

The cerebral hemispheres have four lobes: the frontal lobe, the partial lobe, the temporal lobe, and the occipital lobe. The *frontal lobe* regulates executive functions, planning, and decision making. In other words, the frontal lobe is the CEO. The *parietal lobe* is responsible for reading, attention, short-term memory, spatial awareness, and visual perception. The *temporal lobe* regulates hearing, language, memory, smell, and taste. The *occipital lobe* is responsible for visual processing. The brain can be further divided up into regions and subregions that all participate in a common function based on which types of cells are present in the regions or subregions.

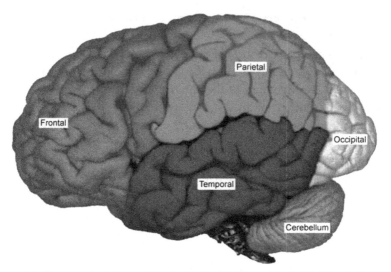

Figure 1.2 The 4 main lobes of the brain: Frontal (purple), Parietal (red), Temporal (blue) and Occipital (yellow).

1.5.3 Pack It All in There

So many brain regions and so little skull space. In order to pack more brain cells into each of those lobes, the brain evolved with folds. Think about packing a suitcase. The suitcase is a certain size no matter what, like the skull. If clothes are simply thrown into the suitcase, it may hold five outfits. But, if the clothes are folded neatly, the suitcase now holds 10 outfits. The brain is the same way. The brain has folds in the tissue to pack in more cells. The top of the folds are called gyri and the grooves or valleys are called sulci or fissures. Based on size, some of the gyri or sulci have names.

1.5.4 Protection: Bones and CSF

We have discussed how important the various steps of brain development are and how necessary proper development is for proper organization and function of the brain. With all of that time and energy invested into making a brain, how does the body protect the brain? The human body protects the brain with bones and with cerebral spinal fluid (CSF- each letter said individually).

The skull or cranium protects the brain. The vertebral column protects the spinal cord. The vertebral column is made of 24 rings like pieces of bone with pliable disks located between the bones. This structure allows us a broad range of movement. If we simply had an inflexible bone instead of vertebrae, there would be no sports. The segments are divided based on function. At the top, near the base of the skull are seven cervical vertebrae known as C1—C7. The thoracic region has 12 vertebrae (T1—T12), the lumbar region has 5 vertebrae (L1—L5) and finally the sacral region has 5 vertebrae (S1—S5).

The brain is floating in CSF. This is a clear fluid that occupies 150 mL of the total volume in the skull (which is 1650 mL). CSF is a filtrate of blood serum. The CSF makes the brain buoyant and reduces the force the brain experiences during acceleration and decelerations by a factor of 30. Liquid can distribute force. So CSF can distribute force that impacts the brain, which further protects the brain.

All of the organizations and specializations discussed in this section occur at early stages of development and persists throughout life which again highlight the necessity of a highly regulated system for proper brain development. Because the brain is so highly regulated during development and because it is necessary for very basic functions, it is protected by bones and the CSF. What can happen when the brain is not protected or when development isn't according to plan?

1.6 WHAT HAPPENS WHEN THE BRAIN IS DAMAGED OR DEVELOPMENT DOESN'T GO ACCORDING TO THE PLAN?

1.6.1 What Happens When the Brain Is Damaged?

1.6.1.1 Spinal Cord Injury and Traumatic Brain Injury

Spinal cord injury (SCI—each letter said individually) research remained very minimal until famous actor Christopher Reeve (1952—2004) who played the role of Superman in the 1978 film, Superman suffered an equestrian accident in 1995. His injury shattered his C1 and C2 vertebrae. Because of this, he was paralyzed from the neck down and was unable to breathe on his own. His abilities despite the severity of the SCI and his determination to find improvements for patients with SCI created a platform for Reeve to spur on research in SCI through his own foundation, the Christopher Reeve Foundation and his co-founded Reeve-Irvine Research Center.

Traumatic brain injuries (TBIs—each letter said individually) and SCI range from mild to very severe. These types of injuries can occur to anyone at anytime. However, there is an increased risk among adolescence because of an increase in risk-taking behaviors, car accidents, and sports-related injuries. With these injuries, a few things happen on the cellular level. Either the sending structure (axon) of the thinking cell (neuron) is cut partially or fully, or the cell body is damaged or the force compresses part of the brain or spinal cord, which can lead to swelling and death of thinking cells (neurons). These injuries can impact blood vessels as well, which can alter how much glucose and oxygen are getting to the nervous system. These injuries are divided into two temporal phases—the primary injury or the force that impacts the head or spinal cord and the secondary phase which is the death of the surrounding cells or blood vessels near the primary injury site. The secondary phase is where current research is targeting approaches to minimize loss of cells around the injury site.

Researchers are exploring what happens to the brain during a traumatic injury. They have performed autopsies on boxers and football players. Interestingly, they have found an accumulation of two proteins that are known to accumulate in the brains of patients with Alzheimer's disease—tau fibrillary tangles and amyloid beta deposits. How these two proteins associated with Alzheimer's as well as TBI remains unknown.

The initial findings of how brain trauma was negatively impacting the brain of professional football players were ignored. In 2002, Dr. Bennet Omalu was assigned the autopsy of Mike Webster, a center for the Pittsburgh Steelers. Dr. Omalu reported a type of TBI, chronic traumatic encephalopathy. Initially, the NFL had league doctors who attacked Omalu's findings. Then, league doctors said that the damage observed was not through football. Finally, in July 2011, the NFL changed its "return to play" policies. To date, researchers with the Department of Veterans Affairs and Boston University have identified chronic traumatic encephalopathy in 131 out of 165 individuals who played competitive football in high school, college, or professionally before their death.

1.6.1.2 Shaken Baby Syndrome

Shaken baby syndrome is a form of TBI or SCI brought about by forcibly shaking a baby or toddler, which can result in permanent brain damage or death. Because the brain and the spinal cord are so delicate, children and babies should never be shaken. In the mid-1800s, forensics experts noted that babies and small children were presenting with nonaccidental brain and SCI. An entire century later, legal action against such injuries became more routine.

1.6.2 What Happens When Development Does Not Go According to Plan?

When development does not follow all the strict guidelines, a wide variety of diseases result. All diseases that affect the brain are categorized based on when things go wrong. Neurodevelopment diseases or disorders are a result of failure of proper development. There are many different developmental disorders. In this section, we focus on three of the more common disorders occurring globally: Down syndrome, schizophrenia, and autism spectrum disorders.

1.6.2.1 Down Syndrome

Up Syndrome (2000) is a critically acclaimed documentary directed by Duanne Graves, who chronicles a year in the life of a friend, Rene Moreno, who is diagnosed with Down syndrome.

There is a known genetic cause for Down syndrome. There is an extra chromosome—a third chromosome 21. Because it has a known genetic cause, Down syndrome is detectable as early as prenatal ultrasounds and can be diagnosed by a genetic screen at or before birth. At the very beginning of development, there was the egg and the sperm, called gamete cells. These cells should have half (haploid number) of the total number of chromosomes that exist in a normal cell (somatic cells). So egg or sperm have 23 chromosomes for a human, whereas a normal cell (somatic cell) will have 46 chromosomes. The process that generates egg or sperm cells (gamete cells) is called meiosis. During this process, cells not only divide, but also they ensure that the egg or sperm (gamete cell) only has 23 chromosomes. But sometimes, the machinery necessary to regulate the chromosome numbers in an egg or sperm isn't working as it should. And an extra chromosome makes it into the egg or sperm. Then the egg and sperm join, and if one of those gamete cells has an extra copy of chromosome 21, that will get passed down to every cell in the body because, remember, once those egg and sperm join, they will start multiplying. Because this disorder is a result from a very early step in development, Down syndrome impacts all cells in the body, not just the brain. Well, mostly. Some people only have a partial triplication of chromosome 21 not a full triplication—but that goes into more science than we will cover here.

An extra copy of chromosome 21 results in decreased muscle tone, flattened nose, smaller facial features, heart defects, digestive problems, hip problems, delayed growth, and problems with obesity. With the brain, people with Down syndrome usually have a lower IQ and delayed cognitive abilities. There are known changes in brains of people with Down syndrome. For instance, some of the structures on the thinking cells (neurons) that form connections with neighboring cells are less stable and there are less of them, which means changes in the way the thinking cells (neurons) communicate. This can lead to problems with learning and memory. There is also a reduction in the size of the brain of people with Down syndrome compared to a nondisease state. Finally, there are alterations in the support cells (glia), which can alter the way that the brain functions. Depending on where these size alterations occur would determine which symptoms this could be responsible for.

There are regions of chromosome 21 that are critical for the hallmark symptoms of Down syndrome. To that end, one of the mouse models used to study Down syndrome has triplicate copies of 150 of the genes found on human chromosome 21 (in the mouse, most of these genes are actually on mouse chromosome 16). Studying these 150 genes will help scientists understand which genes impact the disease most and potentially, how to correct those genes.

1.6.2.2 Schizophrenia

Beautiful Mind (2001) is a movie describing the life of a famous mathematician and Nobel Laureate, John Nash Jr., who was well known not only for his academic achievements but also his battle with schizophrenia. His son, John Charles Martin Nash III, is also a mathematician and is also diagnosed with schizophrenia.

For schizophrenia, there is no one genetic mutation that results in the disorder. There are several types of schizophrenia based on which symptoms are most prevalent. And there are several types of symptoms: positive, negative, and cognitive. The terms positive and negative are based on the effects of the symptoms. The terms are used to describe if the symptoms are lacking, but should be present (negative) or if the symptoms are present but should not be present (positive). Negative symptoms include a loss of motivation, or apathy, loss of normal social behavior, poor use or poor understanding of speech. Positive symptoms include hallucinations, delusions, or paranoia. Cognitive symptoms include impaired memory, disconnected thinking, and impaired executive function. To be diagnosed with schizophrenia, a person must exhibit several of these signs for sustained amounts of time (several months at least).

Ongoing studies are trying to determine if schizophrenia is a result of changes that have been observed: changes in size of brain structures, changes in the survival rate of certain types of cells in regions impacted by schizophrenia, changes in the structure of thinking cells (neurons), changes in the chemical communication (neurotransmission) between thinking cells (neurons) or if it is a cellular change in how the cell moves things from one end of the cell to the other. The other details scientists are trying to figure out is exactly when during development things go wrong and why we don't see symptoms until later in development. Scientists are using scans of human brains and a mouse model that can examine aspects of the disorder, but because there is no single genetic mutation that results in the disorder, mouse models can examine some of the known genetic changes in people with schizophrenia. Piecing together information from various mouse models will help yield an overall picture of what is going on the brain of a patient with schizophrenia.

1.6.2.3 Autism

Mozart and the Whale (2005) is a movie that looks into the life of two people with Asperger's syndrome, one of the disorders that falls under the autism spectrum.

Like schizophrenia, there is no one genetic mutation that results in autism spectrum disorder (ASD). It is thought to be a mixture of genetic and environmental causes. Environmental causes include urban living, paternal age, or a mother who uses valproate to control seizures.

What has been ruled out as a cause of autism are vaccines. There was an original study in 1998 that said there was a link between the measles, mumps, and rubella (MMR -each letter said individually) vaccine and autism. However, that report *was retracted* and viewed by the entire scientific community as fraudulent. In 2012, the Cochrane Library conducted a review of the scientific literature which involved *over 14 million children*, and found no evidence linking MMR and autism. Many other studies looking at large and diverse populations have ruled out vaccines as a cause of any neurodevelopmental disorder. In short, vaccines do not cause diseases. They prevent them. There is a rise in cases of preventable diseases because of not vaccinating children, resulting in long-lasting problems in children due to the disease, including physical and mental disabilities. Serious problems also arise when an unvaccinated, infected child comes in contact with babies who are too young to be vaccinated.

Back to autism. Autisms are in fact a spectrum of several disorders. Most autisms are observed in the first 3 years of life. ASDs are disorders with alterations in three-core symptoms: social interaction, verbal communication, and nonverbal communication. Most people with autism have delayed social abilities, decreased cognitive abilities, and restrictive repetitive behaviors. Because this is a spectrum of disorders, people exhibit a wide range of other symptoms including sensitivity to touch or sound, increased abilities in math or engineering, obsessive compulsive disorder (OCD) tendencies, etc. Medical problems can include sleep disorder, seizures, and digestive problems.

Ongoing studies are trying to determine if some of the observable symptoms in autisms are due to some changes which have been observed in the disease state such as changes in brain region size, changes in thinking cell (neuron) structure in areas of the brain that impact learning and memory, or changes in the balance of cell types in different regions. Many of these changes in the connectivity are localized to the prefrontal cortex, which impacts the social cognition, and perception. Studies are being done in brain scans of people with autism as well as several mouse models. Again, because there is no single genetic cause related to autism, mouse models examine known genetic alterations in people with autism to help provide information about what is different about a brain with autism.

1.7 CONCLUSIONS

It has not yet been determined whether or not the brain and the mind are the same thing or separate entites. The brain, however, is tissue protected by the skull and CSF to generate behavior. Development of the human brain contains a myraid of intricate steps requiring many cells and proteins in each cell to be at the right place at the right time and in the right quantity. Development of the brain can be summarized in five basic steps: (1) creation and closing of the neural tube (neurulation), (2) birth and growth of cells that could become a lot of things (precursor cells), (3) cells find their proper location (migration), (4) precursor cells figure out what kind of cell they are going to be (differentiation), and (5) connections between different cells are made—the brain is starting to be wired together (targeting/synaptogenesis). The brain is undergoing development during childhood through various experiences with the environment and learning and memory tasks that are occurring within the first 7 years of life. Finally, there are several disorders which arise when the brain is not protected such as SCI and TBI. There are also neurodevelopmental disorders in which there are a misstep in at least one of the parts of brain development. Among those are Down syndrome, schizophrenia, and autism.

Why does neurodevelopment matter to society? How can this knowledge inform societies behavior? The intricate dance of neurodevelopment highlights the necessity of good maternal care during pregnancy including educational outreach to potential mothers about why healthy practices during pregnancy are so vital to healthy brain development during pregnancy. The wiring of the brain after birth highlights the necessity for early interventions in any disorder. If any disorder is detected during this time frame, corrective therapies can help in rewiring the brain because of how much wiring is being done during childhood. The brain wiring during these childhood years also highlights the necessity for true experiential learning and memory to occur during this time.

BIBLIOGRAPHY

Alberts, B., Johnson, A., Lewis, J., Raff, M., Roberts, K., Walter, P., 2002. Molecular Biology of the Cell, fourth ed Garland Science, New York, NY.

American Psychiatric Association, 2013. *Diagnostic and Statistical Manual of Mental Disorders*, fifth ed.

Baars, B.J., Gage, N.M., 2012. Fundamentals of Cognitive Neuroscience. Elsevier Academic Press, Amsterdam.

Clark, D.P., Pazdernik, N.J., 2013. *Molecular Biology*, second ed Elsevier Academic Press, MA, USA, Oxford, UK.

Kandal, E.R., Schwartz, J.H., Jessel, T.M., 2000. Principles of Neural Science, fourth ed McGraw-Hill Companies.

Kolb, B., Whishaw, I.Q., 2014. An Introduction to Brain and Behavior, fourth ed. Worth Publishing.

Mason, P., 2011. Medical Neurobiology. Oxford University Press, New York, NY.

Nicholls, J.G., Martin, A.R., Fuchs, P.A., Brown, D.A., Diamond, M.E., Weisblat, D., 2012. From Neuron to Brain, fifth ed Sinauer Associates, Incorporated.

Purves, D., 2012. Neuroscience, fifth ed Sinauer Associates, Incorporated, Sunderland, MA.

Sanes, D.H., Reh, T.A., Harris, W.A., 2012. Development of the Nervous System, third ed Elsevier Academic Press.

Sontheimer, H., 2015. Diseases of the Nervous System. Elsevier Academic Press, London, UK, Waltham, MA, Oxford, UK.

Squire, L., Berg, D., Bloom, F., du Lac, S., Ghosh, A., Spitzer, N., 2012. Fundamental Neuroscience, fourth ed Elsevier Academic Press.

CHAPTER 2

6:00 a.m. Time to Start the Day! How Our Senses Help Us Wake up!

Contents

SUMMARY

The alarm clock is ringing. Several things have to happen at this point to even know that the alarm clock is ringing. How does the brain hear the alarm clock? Cells. Cells in the nervous system interpret signals and information from the environment to give the brain information about the environment and generate behavior based on that information.

Neuroscience Basics.
DOI: http://dx.doi.org/10.1016/B978-0-12-811016-4.00002-7

This chapter will first review what a cell is, then take a brief look at the history of cell theory. This will give us the foundation necessary to understand the two main types of cells in the brain—the support cells (glia) and the thinking cells (neurons). Understanding what a cell is and which cells make up the brain, we can better understand how the brain uses those cells to perceive the world around us. Finally, this chapter will look at what happens when cells of the brain do not function properly.

2.1 CELLS 101

Humans are made up of cells. And cells are what allow us to determine the alarm clock is ringing and what to do about it. Here are some basic structures and functions of the cell by way of reminder to better understand how cells work within the brain.

First, cells contain the genetic information. Remember that genes are made up of DNAs that contain a code for a certain protein. Genes made of DNA codes are located in a safe place inside the cells called the nucleus. During a decoding process called transcription, DNA is read and then makes an intermediate code, mRNA. Then, during another decoding process (translation), the mRNA code is read to create a protein. The need for the two steps of the process are in part due to the fact that the DNA is stored in a vault (the nucleus) that is hard to get access to. This is a good thing because the DNA is the blueprint or the upper level management for cells. The things necessary to make the protein cannot get into the vault so that they don't accidentally mess up the DNA. DNA is protected and cannot leave the vault (nucleus) so that no one messes it up. So an intermediate code, the mRNA, is made in the vault and then leaves the vault so it can be translated into protein (Fig. 2.1).

The DNA is in the nucleus and is decoded into mRNA in the nucleus. Then the mRNA leaves the nucleus and heads to the endoplasmic reticulum (ER) and ribosomes. The ER and the ribosomes both play a role in translating proteins from mRNA and folding proteins after they have been translated. Proteins have to be folded properly in order for them to function. Think of protein folding like folding a parachute. If the parachute isn't folded and packed properly, it won't work. The same is true for proteins.

After being properly folded, proteins have to get to the part of the cell where they can carry out their function. The Golgi is like a post office of sorts. The Golgi prepares proteins to be shipped throughout the cell and makes sure they have the appropriate address labels. Once a protein has

GENERALIZED ANIMAL CELL

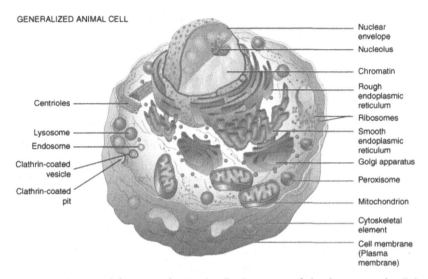

Figure 2.1 Structural features of animal cells. Summary of the functions of cellular organelles. *Mitochondria*: (1) Site of the Krebs (citric acid) cycle; produce ATP by oxidative phosphorylation. (2) Can release apoptosis-initiating proteins, such as cytochrome c. *Cytoskeleton*: Made up of microfilaments, intermediate filaments, and microtubules; governs cell movement and shape. *Centrioles*: Components of the microtubule organizing center. *Plasma membrane*: Consists of a lipid bilayer and associated proteins. *Nucleus*: Contains chromatin (DNA and associated proteins), gene-regulatory proteins, and enzymes for RNA synthesis and processing. *Nucleolus*: The site of ribosome RNA synthesis and ribosome assembly. *Ribosomes*: Sites of protein synthesis. *Rough ER, Golgi apparatus, and transport vesicles*: Synthesize and process membrane proteins and export proteins. *Smooth ER*: Synthesizes lipids and, in liver cells, detoxifies cells. *Lumen*: Ca^{2+} reservoir. *Clathrin-coated pits, clathrin-coated vesicles, early and late endosomes*: Sites for uptake of extracellular proteins and associated cargo for delivery to lysosomes. *Lysosomes*: Contain digestive enzymes. *Peroxisomes*: Cause β-oxidation of certain lipids (e.g., very long chains of fatty acids). *Modified from Freeman, S., 2002. Biological Science, first ed. Prentice Hall, Upper Saddle River, NJ.*

been used or if the protein is misfolded, it becomes waste. The lysosome breaks down waste in the cell.

There are a few other structures that are necessary for the cell. The cell has support beams and transport highways. Microfilaments and tubules act as transport highways and support beams throughout the cell. The intracellular fluid and the cell membrane give the cell its shape. All of the cell's functions require energy. The powerhouse for the cell that stores and releases energy is the mitochondria.

All of these parts that make up a cell are found in brain cells as well. Each of the functions for the parts of the cell are necessary for the brain to carry out functions, like hearing an alarm clock go off. But before we get more into the function of brain cells, let's look back at how we know that the brain is made up of cells.

2.2 HISTORY OF CELL THEORY

The mid and late 1800s were an amazing time for science. Remember in our previous chapter, it was the mid to late 1800's where scientists like Paul Broca were determining that regions of the brain were responsible for specific behavior. This time period was also important to our understanding of cells. Around the 1860's several prominent scientists such as Matthias Schleiden, Theodor Schwann, and Rudolf Virchow, all contributed to cell theory. Cell theory states that cells are the basic unit of life, that all living organisms are made up of at least one cell, and that all cells arise from a preexisting cell. This was a hotly debated theory in the early 1800s. As low-powered microscopes became available, scientists were able to generate evidence to support cell theory.

Cells are no longer the area of hot debate. Today, cells are taught as early as the first or second grade. In fact, one of the popular assignments in elementary school is constructing a cell for science class. Some teachers prefer the edible cell project—a brownie covered in candies all labeled with cell part names.

Once cell theory had been established at the organismal level, a new debate about cells and brain began. In the late 1870s, scientists of the day believed the brain was a mesh of interconnected cells, not individual cells. Camillo Golgi had developed a technique using silver nitrate to stain cells and look at the nervous system. Santiago Ramón y Cajal used Camillo's technique to examine brain cells. What is brilliant about Cajal's work is that he did not have a digital camera hooked up to his microscope as neuroscientists today utilize. He drew brain cells by hand. By hand!!! And this is one of the many junctures where science and art meet. Put down this book for a moment and search "Ramón y Cajal Neurons" on the web. His images are amazing and fascinating. The work of Golgi and Ramón y Cajal settled the debate—the brain was made up of individual cells. Their work earned them a Nobel Prize in 1906.

Each cell in the brain contains organelles, like the ones we discussed in this chapter, which are necessary for proper function. And each cell in the brain is an individual cell. But what types of cells are in the brain? How are they different and what are their roles?

2.3 SUPPORT CELLS

One of the amazing things about the brain is that there at least 100 billion cells in the brain. There are two main types of cells—the "thinking cells" that are responsible for the actions of the brain—the neurons—and the supporter cells that make sure the thinking cells do their job—the glia. If you look at these two types by the amount of space they take up in the brain, the supporting cells (glia) and the thinking cells (neurons) are pretty similar. But if you look at the brain by number of cells, glia out number neurons at least 2:1.

Neurons are the thinking cells. They are the cells receiving and sending information within the brain. But what are the supporting cells (glia)? What do they do? Initially, there was very little research on support cells (glia) because their role in the brain was thought to be minimal. After all, they did not do the communicating. However, research has demonstrated massively important roles for these support cells (glia) in the brain. Thinking cells (neurons) alone cannot make a brain. Support cells (glia) are necessary in order for the brain to function.

There are at least five different types of supporting cell (glia) types and they have different purposes.

1. Microglia are a type of supporting cell (glia) found in the brain. Microglia act as the immune defense in the central nervous system (CNS) and make up roughly 10−15% of the total cell count in the brain.
2. Oligodendrocytes (from the Greek meaning "cells with a few branches") or oligodendroglia (from the Greek meaning "few tree glue") are supporting cell (glia) found in the brain. Oligodendrocytes provide support and insulation to the sending structure (axons) on thinking cells (neurons) in the brain. Oligodendrocytes wrap a myelin sheath, which is 80% fat and give the white appearance to white matter. The myelin sheath that is wrapped around the sending structure (axon) also aids in quick communication.
3. Schwann cells wrap the sending structure (axon) of thinking cells (neurons) in the peripheral nervous system (PNS).
4. Astrocytes (from the Greek meaning "star cell", which describes their structure) are supporting cells (glia) in the brain. They perform many functions, including formation of the blood−brain barrier, providing nutrients to the brain, balancing the ions necessary for brain communication, and repair and scar formation following injury.
5. Radial glia are supporting cells (glia) found within the brain. During development, they guid cells from their birthplace to the region of the

brain where they will operate. Radial glia are like a GPS for thinking cells (neurons) during development.

While the supporting cells (glia) are necessary for proper brain function, this chapter is going to focus on the thinking cells (neurons).

2.4 THINKING CELLS (NEURONS)

The tight relationship between structure and function is an underlying theme in almost all of biology. Thinking cells (neurons) have a few key structural components. Each structure helps the thinking cell perform a function.

Interestingly, the structure and function of thinking cells (neurons) are very similar to skin cells (epithelial cells), which makes sense based on development. The nervous system is generated from the endoderm, which is the same layer (epiblast) that the skin is made from. An example of their structural similarity—both skin cells (epithelial cells) and thinking cells (neurons) have distinct "this end up" regions (called poles).

Thinking cells have structures that receive messages (dendrites) that look like tree branches. There is a cell body that contains a lot of the compartments that organize and make proteins and a sending structure that sends the message (the axon and the axon terminal) (Fig. 2.2).

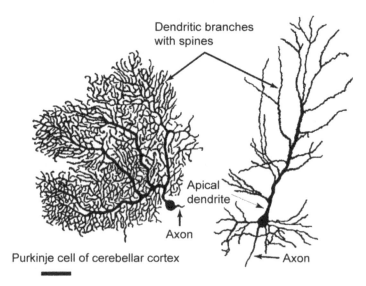

Figure 2.2 Two types of neurons within the brain with unique dendritic and axonal structures.

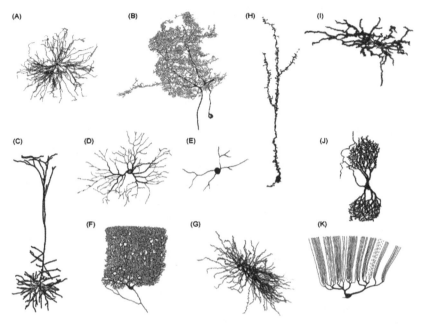

Figure 2.3 Variation in the structure of dendritic trees. Reconstructions are shown from: (A) alpha motoneuron of cat spinal cord, (B) spiking interneuron from meso-thoracic ganglion of locust, (C) neocortical layer 5 pyramidal neuron of rat, (D) retinal ganglion cell in cat, (E) amacrine cell from retina of larvar salamander, (F) cerebellar Purkinje cell in human, (G) relay neuron is basoventral thalamus of rat, (H) granule cell from olfactory bulb of mouse, (I) spiny projection neuron from striatum of rat, (J) nerve cell in nucleus of Burdach of human fetus, and (K) Purkinje cell of mormyrid fish.

Specific thinking cells (neurons) adapt these structures for their own purpose within the brain. Cells in the retina have less bushy receiving structures (dendrites) than cells found in the brain region for learning and memory or in the brain region for movement. Cell bodies can have different shapes and the sending structure, the axon, can have different lengths and shapes all depending on the brain region. The structure of each individual thinking cell (neuron) depends on what its function is within a given brain region (Fig. 2.3).

The brain is made up of cells that have various tasks parts to carry out specific functions. Each of these individual units of the brain are either a supporting cell (glia) or a thinking cell (neuron). How do these cells help us start off our day? How do cells interpret information gathered by our senses?

2.5 THE SENSES AND THEIR CELLS

All of the sensory information we receive from our environment are first sensed by our primary sensation organs—ears, nose, tongue, eyes, and touch could be any part of our body with functioning nerve endings. The sensory information is received by a cell. That cell translates the information/stimulus into a signal that it sends to the brain region specific for that sensation (Fig. 2.4). The brain initially receives the information, and then based on previous experiences, it interprets the information. This interpretation of the sensation that is experienced is called our perception. And because we all interpret differently based on how different our sensation may be and how different our experiences may be, our perception of the environment around us is subjective.

There are a few types of cells that sense the information/stimulus in the environment. Mechanoreceptors receive information through stretch, touch, or movement of parts of the cell. Chemoreceptors receive the information through a chemical stimulus (smell or taste).

Information from the sensing thinking cell (sensory neuron), which can be found all over the human body, conveys that information to the

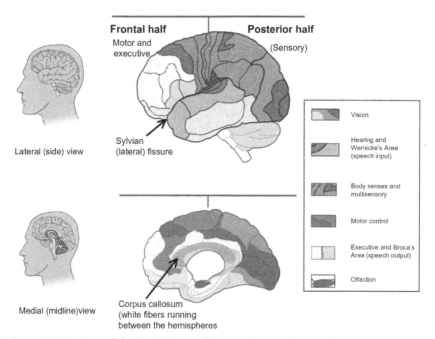

Figure 2.4 Regions of the brain involved in sensation and perception.

somatosensory cortex. The somatosensory cortex forms a homoculus, a little man, when the areas of sensation are mapped onto the cortex.

Back to the alarm clock and the fact that we are starting our day.

2.5.1 6:00 a.m.—The Alarm Clock Goes Off

Question—How do cells hear the alarm clock?

Answer—Sounds like an alarm clock are detected because of a type of cell— hair cells!

The ear has special cells in a part of the ear called the cochlea. The special cells are the hair cells (named because of their appearance). These cells are suspended in a fluid. The fluid moves when hit by sounds waves or by the head moving, creating movement of the hair cells. The hair cells of the cochlea have an acoustic frequency range from 20 to 20,000 Hz. The hair cells in the semicircular canal are stimulated by angular acceleration during head rotation. The top of the hair cell is referred to as the hair bundle. The hair bundle is very sensitive to movement and with the slightest of movement caused by small sounds or small head motion, the hair bundle will shift. This allows channels in the membrane of the hair cell to open creating a shift in ions, which in turn generates an action potential—a change in electricity that the brain cells will use as a message.

The movement of the hair cells tells the brain a lot of different types of information about the sound (intensity, duration, pitch). This information is then translated into signals that are shipped off to the auditory cortex in the brain.

One of the key functions of sound is to interpret words being spoken, or language. Initially, it was thought that both the left and right hemispheres of the brain did the same thing. But in the mid-1800's, research by French physician Paul Broca began to question that theory. Broca examined several postmortem brains and concluded that language was localized to the left hemisphere. The area of the brain now known as Broca's area, is the language center of the brain. Later, another researcher described a speech region of the brain, named after him, Wernicke's area. The difference between the two areas: Broca's area stores motor programs for speaking words, while Wernicke's area contains sound images of words.

While interpreting sound also involves learning and memory, the basics of hearing that alarm clock are the hair cells receiving the input of sound, sending a message to the thinking cells (neurons) in the auditory cortex in the brain, the thinking cells (neurons) in auditory cortex telling the thinking cells (neurons) in the decision-making areas of the brain that the sound heard is an alarm clock. The thinking cells are in constant communication!

2.5.2 6:01 a.m.—Time to Hit the Snooze Button

Question—How does the brain know where the alarm clock is and where the button is to press the snooze?

Answer—The brain can detect how hard and where to press a button because of cells—sensory thinking cells (sensory neurons)

The human body is covered with cells that measure sensation both on the outside and on the inside. The density of these sensing cells (sensory neurons) is different in different parts of the body. That is why it may hurt to get a tattoo on one part of your body versus very little pain in another location. This is also part of the reason why people may have different pain thresholds—they may have a different density of sensing cells (sensory neurons).

These sensing thinking cells (sensory neurons) can detect (among many other things)

Hapsis: the ability to differentiate objects based on touch, fine touch, and pressure

Nocioception: pain, excess heat, pressure, chemicals, tissue damage

Thermoreception: temperature of the skin and blood

Photoreception: light detection.

Pain is necessary. Pain drives us to correct things that are harmful, motivates us to protect our body from certain situations, and urges us to avoid similar experiences in the future. Pain is a perception that is created through a lot of sensory information. It takes information from touch and pressure (Hapsis) and proprioception (body location and movement) and sends that information to the brain so the brain can figure out what to do about the pain. Pain is either acute (short lasting) or chronic (long lasting). Pain can be mild or it can be severe. It can be steady, throbbing, stabbing, pinching, aching, etc. The type of pain, the severity, and the time period of the pain all depend on the situation and what is causing the pain. Pain is the result of information from all of these sensory thinking cells (neurons).

There is another system that combines information from several sensory thinking cells (neurons)—proprioception. The body has to sense where it is within the environment. Proprioception is the information that is sent back to the brain about body location and body movement. This information aids in our movement and our balance. Back to the ear we go to understand proprioception. In the ear is a system called the vestibular system. This system is responsible for telling the brain where the body is in relation to gravity and this systems tells the brain the direction and speed of head movements. This system, like the part of the ear needed for hearing sound, also has fluid and hair cells. When the head moves, the fluid moves, which moves the hair cells. This movement of the hair cells tells the brain how fast the head is moving and in which direction. All of this information tells the brain the location of the body in space.

Let's play with this for a minute. Put a hand in front of your eyes and shake it vigorously. It seems blurry, right? Now, hold that hand still and shake your head back and forth. The hand remains in pretty good focus, right? This is because your vestibular system sent signals to your brain allowing you to compensate for the head movement.

When proprioception doesn't work, we experience vertigo (from the Latin for "spinning"). Vertigo is a sensation of dizziness when you aren't actually moving, and can be paired with nausea and/or inability to balance while walking or standing. If you want to try this for fun, spin around and around in a circle for about a minute and then try to walk a straight line!

Back to hitting the snooze button. We are able to detect how much pressure we should apply on the snooze button because of haptic cells. We are able to know where our hand is in relation to the snooze button because of proprioception. All of the sensing cells (sensory neurons) are sending information about where our hand is and where the alarm is and how hard to press the snooze button. This information moves from the sensing cells (sensory neurons) to the thinking cells (neurons) in the somatosensory cortex which then sends the information to the thinking cells (neurons) in the decision-making area of the brain.

2.5.3 6:15 a.m.—Time to Find the Kitchen

Question—How do humans see where they are going?

Answer—Rod and Cone cells detect information and the brain decodes that information

The snooze alarm goes off and it is finally time to get out of bed. When your eyes finally stay open, how do they see? How do they detect the environment around you to keep you from walking into a wall (most of the time) (Fig. 2.5).

Light travels anywhere from the outside world. It enters our pupils and goes to the back of the eye to the retina. In the retina, there is a region called the fovea. The fovea collects the majority of the light. And it does this upside down. The information collected by the eye, the images that it is picking up, are projected into the eye upside down. Weird fact—if you

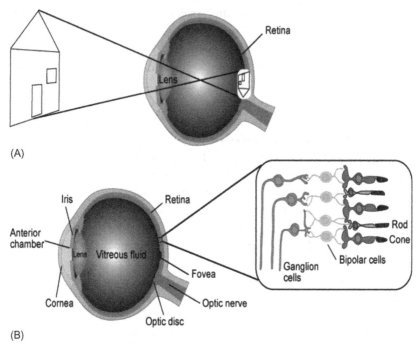

(A)

(B)

Figure 2.5 Cells of the Eye — Rods and Cones.

were to put on glasses that inverted the world, after several days, the world would appear right side up again because the brain would correct it.

In the retina at the back of the eye, there are special cells that are receiving the information about the light. These cells were named based on their shape—they are called rods and cones. There are more rods and they are sensitive to levels of brightness. Cones respond to bright light, mediate color vision, and our ability to see fine detail. The fovea has only cones, which is why we look at something directly when we want to focus. Our peripheral does not have as many cones, which is why our peripheral vision is not as strong as looking straight on.

Information from the rod and cones travels to the thinking cells (neurons) in different parts of the brain. Some of the information is sent to the hypothalamus, which helps regulate the circadian rhythms. Other information is sent to the pretectum, which controls the reflex of the pupil and the lens. And other information is sent to the superior colliculus for orienting the movements of the head and the eye. Other information is sent to regions of the brain to decode the objects we are seeing.

This information does a few things for us; it tells you what is around us but this information also tells us about our location (depth as well as the objects around us). The brain constructs maps so we don't run into things. These maps are created based on which eye the information came from and where on the retina that information hit.

With a solid spatial map of the room, we are able to stumble out of bed and head to the kitchen to start our day.

2.5.4 6:20 a.m.—The Coffee Machine Turns On and the Aroma of Coffee Fills the House! *Must Drink Coffee*!

Question—How does the brain know that the smell in the air is actually coffee?

Answer—Cells detect chemical scents!

Smell (olfaction) is a complex sense. Mammals use approximately 100,000 olfactory receptor thinking cells (neurons) to detect odors. We have the ability to detect and differentiate between a wide variety of odors (approximately 450 different types), but we often don't have words for them all.

Scents are chemicals and olfactory thinking cells (neurons) are chemoreceptors. Chemicals bind to the olfactory thinking cells (neurons) and open channels in the membrane causing ions to move, which activates messengers within the cell (signaling cascade pathway). These messages get passed from the chemoreceptor up the chain. These olfactory thinking cells (neurons) send their information through a thin portion of the skull (the cribriform plate) to the olfactory bulb, which then communicates with the thinking cells (neurons) of several regions of the brain. Some information is sent to the orbitofrontal cortex (to integrate the smell with taste information), the thalamus (to relay to other regions of the brain), the hypothalamus (emotional response) and the hippocampus (for learning and memory). Because these different brain regions are involved and because of the response they elicit, smells are tied to many different types of responses. For example, if you smell chocolate chip cookies, your body may feel hungry and cannot wait to eat a cookie (orbitofrontal cortex) and you may remember a time in your childhood making chocolate chip cookies with a loved one (initially sent from the hippocampus to the cortex) and you have an emotional response to remembering that experience (initially sent from the hypothalamus to the cortex to be stored as a memory). Smell is not only the ability to sense odor but also strongly linked to learning and memory (Fig. 2.6).

The olfactory thinking cells are located in the nasal cavity, which is covered by mucus that has to be replaced every 10 minutes. The mucus serves as a barrier in the nasal cavity. It protects the olfactory thinking

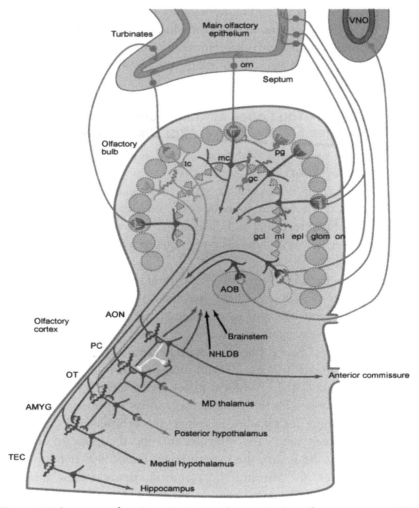

Figure 2.6 Summary of main projection pathways in the olfactory system. AON, anterior olfactory nucleus; PC, pyriform cortex; OT, olfactory tubercle; AMYG, amygdale, TEC, transitional entorhinal cortex; NHLDB, nucleus of horizontal limb of diagonal bend; MD, mediodorsal.

cells (neurons) from toxic airborne odors. Olfactory thinking cells (neurons) have a lifespan of $1-2$ months.

The complexity of smell is exemplified in the training required of wine experts. Wine experts rely on smell (olfaction) to tell them about wine. To become wine experts, people go through hours of training in order to train their nose to tell them about a wine correctly.

Question—How does the brain communicate what the coffee tastes like?

Answer—Different types of cells on our tongue determine how the coffee tastes.

Taste is also called gustation. Taste is similar to smell in that it is chemical information being received from the environment. Some of the taste stimuli act on the channels in the membrane, while others act directly on the membrane.

Taste preferences within the human species are diverse. Cells on our tongue are different. Some humans can detect low levels of certain chemicals, so some foods taste very different to them to those who do not detect low levels, only mid-range levels. For example, there is a chemical (6-*n*-propylthiouracil, or PROP—pronounced as a word) that determines bitterness. People who can detect PROP at really low levels are called supertasters. People who do not classify PROP as bitter are called nontasters. For fun, you can try this at home. You can order 100 tests (with control strips with other tastes on them) on Amazon. Nontasters eat more vegetables. And, the ability to detect levels of PROP are genetic (Fig. 2.7).

There are taste receptors all over the mouth and of course on the tongue. There are four main types of receptors—sweet, sour, salty, and bitter. Each taste bud has several types of each type of taste receptor. When a

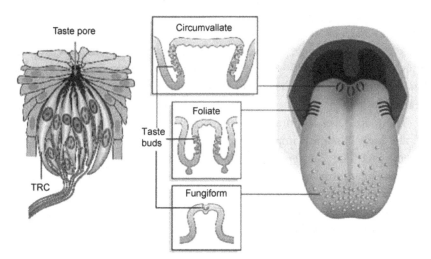

Figure 2.7 Diagram of tongue showing the distribution of various taste bud populations, which are found in the fungiform papillae on the anterior tongue, the vallate and foliate papillae on the posterior tongue. These taste buds are innervated by branches of the VIIth, IXth, and Xth cranial nerves (see text). *Reprinted by permission from Nature. 2006. The receptors and cells for mammalian taste. Chandrashekar J1, Hoon MA, Ryba NJ, Zuker CS, Nov 16;444(7117):288—94.*

receptor detects it's specific "taste," the sensing cell (sensory neuron) on the tongue sends its signal to the thinking cells (neurons) first in the thalamus (relay station), then in the gustatory cortex and then in the hypothalamus (emotional response) and the amygdala (to assist in the food memory).

Taste is different in children than adults. Children have almost twice as many taste receptors than adults so food is very different to children than adults. This is why children's food preferences change as they grow. Their taste buds are literally changing as they grow up.

There are entire classes dedicated to understanding sensation and perception. That is a very basic and brief introduction to our sensation, which is our sensory organs detecting information in the environment around us. Perception is the brain's interpretation of that information based on previous experience. What happens when sensation is altered? With the understanding gained in this chapter about the importance of cells in the brain, what happens when the cells do not function properly?

2.6 WHAT HAPPENS WHEN SENSATION IS ALTERED?

2.6.1 Hearing

There are both hearing impairments and hearing loss. A person with little to no hearing is classified as deaf. Hearing loss may occur in one ear or both and may be temporary or permanent. Loss of hearing may be a result of genetics, aging, exposure to noise, infection, trauma, or medications. Approximately 70–85% of hearing loss is inherited. Some hearing loss is associated with other disorders that impact hearing (syndromic) or hearing loss that is not associated with any other disease (nonsyndromic).

There are several options that can aid the patient with regaining some of the loss. Cochlear implants can improve outcomes for people with hearing loss. They artificially provide the electrical stimulation to the thinking cells (neurons) in the cochlea. Some people with deafness oppose attempts to cure it and view cochlear implants with concern and negativity.

On the more well known documentaries on hearing and vision loss is the story of Helen Keller in *The Miracle Worker* (2000).

2.6.2 Vision

Visual impairment or vision loss is a decrease of the loss of the ability to see. Visual impairment can be corrected in part with glasses or contacts, but vision loss is usually permanent and unable to be corrected. Blindness is a term usually designated for complete or nearly complete vision loss. The most common visual impairments are near sighted, far sighted, cataracts, and glaucoma. Disorders that may cause visual problems are macular degeneration, diabetes, infections, stroke, or trauma.

At First Sight (1999) is a movie that explores the life of a patient who suffered visual loss and regained some vision through surgery.

In addition to visual impairments and visual loss, there are several different types of color vision deficiencies ranging from a particular color weakness to a color blindness. The most common deficiency is the inability to distinguish between red and green (red—green color blind) predominately but also gray and purple as well. Red—green color blindness is a mutation on the X chromosome and therefore is an inherited trait. The mutation results in affecting the pigments found in the cone cells of the retina. Usually it is a genetic condition, but it can result from physical or chemical damage to the eye.

There are other types of color blindness as well, including blue—yellow color blindness and complete color blindness.

2.6.3 Phantom Pain

Phantom pain is experienced by anyone who has lost sensation in particular part of the body, but the patient still experiences "pain" in that region. This is due, in part, to the fact that the brain regions necessary to control that region are still intact. The loss of feedback from that region results in phantom perceptions from that region. To help alleviate chronic phantom pain, local anesthetics are administered to the nerves near the region that has lost sensation.

In particular, phantom limb pain is reported among amputees who report pain, muscle cramps and movements in the limb that has become amputated. It is more widely experienced in people who have had upper limbs amputated as compared to people who have had lower limbs amputated.

2.7 WHAT HAPPENS WHEN CELLS DO NOT FUNCTION PROPERLY?

2.7.1 Amyotrophic Lateral Sclerosis

Cells are living things. And we lose some of them due to natural consequences. But there is a disease that is caused by overdeath of thinking cells (neurons) found in the spinal cord, Amyotrophic Lateral Sclerosis (ALS). ALS is a progressive disorder caused by the death of the motor thinking cells (neurons) found in the brain and the spinal cord. Sensory thinking cells (neurons) remain unaffected by this disorder. It usually affects the motor thinking cells (neurons) found in the motor cortex of the brain (upper motor neurons), which activate the lower motor neurons (thinking cells) found in the spinal cord or brain stem. Because those thinking cells (neurons) die, they cannot tell the muscle what to do, and because of that, the muscle will go unused and eventually result in muscle wasting due to unuse.

Similar to traumatic brain injury and spinal cord injury, much attention was not given to ALS, which was first described in the late 1800s, until it impacted a famous athlete in the mid-1900s—New York Yankee Lou Gehrig. Today ALS is also known as Lou Gehrig's Disease. Gehrig left his baseball career after a diagnosis with ALS. Attention was also given to disease because the brilliant theoretical physicist, Stephen Hawking was diagnosed with the disease at the age of 21.

Mutations in approximately 16 genes are believed to be the cause of both familial and sporadic ALS. Because of this, there are several mouse models of the disorder. The disease is usually observed in people aged 40−50. In ALS, there is excessive death of thinking cells (neurons) that control muscles. The main symptom is muscle weakness in a limited number of muscle groups initially but eventually includes more muscle groups. Initially, the disease affects muscles asymmetrically, but eventually the disease will impact all muscles except the muscles that control eye movement and bowl and bladder function. In people, this usually results in paralysis, difficulty swallowing, and required assistance for breathing and speaking. Life expectancy is usually several years after the diagnosis, but it is different for each case. For example, Stephen Hawking was diagnosed with ALS in 1963 and given only 2 years to live. But thankfully today, he is still writing books and giving lectures. *The Theory of Everything* (2014) is a movie depicting the relationship between Hawking and his wife as he learned of his diagnosis and as the disease progressed.

2.7.2 Glioblastomas

Cells in our body grow and divide, which creates new cells, until there is a switch in the cell that tells it to stop growing and dividing. Cancer occurs when that switch gets ignored and the cell keeps growing and dividing. This can happen in any part of the body. The big problem with cancer in the brain is that we don't have extra space, so tumors end up pushing on areas of the brain that we need for every day functions. There are several types of cancer (lung, breast, skin, or kidney) that may typically metastasize to the brain through blood circulation. The problem with tumors that have metastasized into the brain is that they are usually present at late stages of other forms of cancer which present their own set of issues.

Primary CNS tumors originate and typically remain within the brain. There are several types of primary brain tumors based on the location and what type of cells are involved. Brain tumors caused by unchecked growth of glia cells are called gliomas and they are the most common type of brain tumor. Here we are going to focus on gliomas, which represent the largest percentage of cases of primary brain tumors and include astrocytomas, oligodendrogliomas, and glioblastomas. The only risk factor identified with gliomas is ionizing radiation. Gliomas are more common among the very young and the elderly. The symptoms are highly variable and based on which region of the brain the glioma is located. There are a few common symptoms such as headaches, nausea, brain dysfunction, seizure, personality changes, or muscle weakness. Gliomas are rated on a scale of 1−5 with increasing number signifying higher malignancy. As with many cancers, surgery, radiation, and chemotherapy remain the standard of care, however, they are not necessarily a cure.

2.7.3 Ethical Implications of ALS and Glioblastomas

Interestingly, both ALS and Glioblastomas bring up an ethical question—death with dignity. Death with dignity is an end of life option that is currently legal in Oregon, Washington, and Vermont for mentally competent patients diagnosed by at least two physicians with a terminal illness leading to death within 6 months. Patients who fulfill those requirements undergoing waiting periods in between their first and second request for a prescription that will end the patient's life. Looking at patients who use such a prescription in Oregon, 69% have cancer, including glioblastomas and 16% have ALS. These laws made headlines again with a patient who had glioblastoma, Brittany Maynard. Brittany Maynard was 29 years old when she was diagnosed with a brain tumor. Her intense and heart breaking story made national headlines as she moved to Oregon in order to avail herself of Oregon's death with dignity laws.

There are so many crossroads between science and society and political policy. It is necessary for citizens to have a basic understanding of the biology behind such diseases so that they can make informed decisions when they vote, which impacts our neighbors who are dealing with terminal illnesses.

2.8 CONCLUSION

2.8.1 Overview of the Information From Our Senses

The five senses give the brain information about the environment. All of this information tell the brain about the world. The sensory information also gives feedback on what is going on with the body in the world. Humans perceive the information or detect the information through the senses differently from one another. Human perceptions of reality are different because humans sense things differently and because each human is unique in the way that they may perceive a color or a sound or a sunset, we should respect and embrace these differences in opinions. All of the senses change with experience and change with age. During aging, humans loose some eyesight capabilities and some hearing capabilities as well. The sooner these issues are corrected, the better for our brain because corrected hearing or corrected vision keeps the brain cells talking with each other. Keep your brain sharp no matter what!

BIBLIOGRAPHY

Alberts, B., Johnson, A., Lewis, J., Raff, M., Roberts, K., Walter, P., 2002. Molecular Biology of the Cell, fourth ed. Garland Science.

Augustine, J.R., 2007. *Human Neuroanatomy.* Elsevier Academic Press.

Baars, B., Gage, N., 2012. Fundamentals of Cognitive Neuroscience. Elsevier Academic Press.

Cajal, S.R.Y., 1995. Histology of the Nervous System of Man and Vertebrates. Oxford University Press.

Clark, D.P., Pazdernik, N.J., 2013. Molecular Biology, 2nd Edition Elsevier Academic Press.

Cowan, W.M., Sudhof, T.C., Stevens, C.F., 2001. Synapses. The Johns Hopkins University Press.

Goodman, S.R., 2007. Medical Cell Biology, third ed. Elsevier Academic Press.

<https://www.deathwithdignity.org/faqs/>.

Kandal, E.R., Schwartz, J.H., Jessel, T.M., 2000. Principles of Neural Science, fourth ed. McGraw-Hill Companies.

Kolb, B., Whishaw, I.Q., 2014. An Introduction to Brain and Behavior, fourth ed. Worth Publishing.

Mason, P., 2011. Medical Neurobiology. Oxford University Press.

Nicholls, J.G., Martin, A.R., Fuchs, P.A., Brown, D.A., Diamond, M.E., Weisblat, D., 2012. From Neuron to Brain, fifth ed. Sinauer Associates, Incorporated.

Peters, A., Palay, S.L., Webster, H.D., 1992. The Fine Structure of the Nervous System: Neurons and Their Supporting Cells, third ed. Oxford Press.

Purves, D., 2012. Neuroscience, fifth ed. Sinauer Associates, Incorporated.

Sontheimer, H., 2015. Diseases of the Nervous System. Elsevier Academic Press.

Squire, L., Berg, D., Bloom, F., Du Lac, S., Ghosh, A., Spitzer, N., 2012. Fundamental Neuroscience, fourth ed. Elsevier Academic Press.

CHAPTER 3

6:35 a.m. Time to Run—How Does the Brain Tell Our Muscles to Move?

Contents

SUMMARY

We have had our coffee and we are waking up. Time to go for a run. How does the brain tell the body to move? In order to understand how the muscle cells and the thinking cells (neurons) interact with each other, we have to understand the type of signals that they use to communicate—the action potential. The action potential is a change in electricity in the thinking cell (neuron) because of a change in the concentration of ions—particularly sodium and potassium. The brain keeps those ions out of balance (different amounts inside and outside the thinking cell), and when the conditions are right, they move across the thinking cell's

Neuroscience Basics.
DOI: http://dx.doi.org/10.1016/B978-0-12-811016-4.00003-9

membrane in order to have the same number of ions inside and outside of the cell. When that happens, it changes the electricity in the membrane of the thinking cell (neuron), which generates an action potential, which is how the thinking cell (neuron) talks to the neighboring cells, including the muscle cells. How do we know all this? We will briefly discuss some amazing historical experiments that shed some light on the action potential. What happens with the action potential between the thinking cell (neuron) and the muscles is not clear or is not sent? There are several movement disorders that are a result of a failure of the brain's thinking cells (neurons) to send appropriate messages to the muscles and we will review a few of those disorders in this chapter.

3.1 HOW DOES THE BRAIN MOVE THE BODY?

Exercise does a lot for the brain, including reducing stress! Our brain has tight regulation over our movements, allowing us to make complex movements.

3.1.1 How Does the Brain Generate the Commands for Physical Activity?

Many people engage in some sort of physical activity each day ranging from walking, wheeling, or training for intense competition. For all of these activities, the brain must first generate a command, which is sent to the muscles through nerves in the spinal cord, which move the body in the direction and intensity directed by the brain (Fig. 3.1).

Initially, a part of brain that is necessary for planning and initiating motor sequences (prefrontal cortex) is activated. Then, a region that

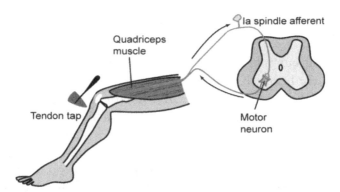

Figure 3.1 The knee-jerk reflex is an example of a spinal cord reflex. The communication between the brain and the muscles shows all the connections necessary to generate movement.

produces the complex movement sequences (premotor cortex) is activated. Finally, information is sent to a region that tells the body how each movement is to be carried out (primary motor cortex). The muscles send back information about where they are in relation to the world around them (proprioception) as well as other necessary information to further generate movement (Figs. 3.2 and 3.3). The motor cortex has a map of areas that each part of the motor cortex is responsible for. This map makes up a homunculus—little person, just like the sensory cortex.

The more complex the movement, the more active your brain is. If the movement is exercise you do frequently, your brain will learn the movement and end up going to "autopilot" to generate the movement, however, you will still have to generate motivation to continue the movement. Some exercise or movement may require more motivation than others! Even though you are on "autopilot", your brain is still active, recovering the memories of the movement and making calculations about force and directionallity.

Movement also requires two other areas of the brain. One area (basal ganglia) controls force of movement and computes the effort costs necessary in making movement. The other area is an entire region of the

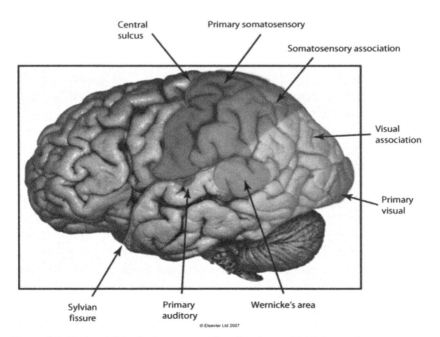

Figure 3.2 Regions of the brain necessary for planning and initiating motor sequences.

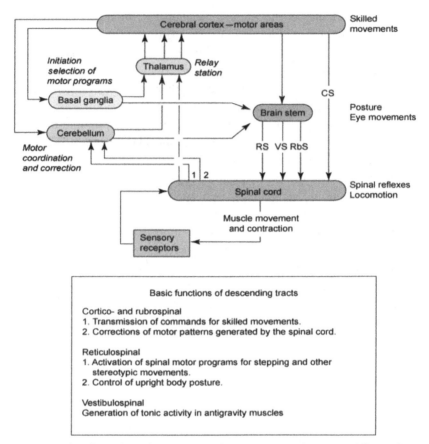

Figure 3.3 This flow chart demonstrates the connections between different brain regions necessary to generate motor sequences.

brain dedicated to movement—the cerebellum. The cerebellum is necessary for timing and execution of movement. The cerebellum is divided into several regions, each of which regulates motor control.

Movement that is repetitive, or produces a rhythm, can be controlled by a group of thinking cells (neurons) called central pattern generators. Even though they produce rhythmic movements, they do not control dancing. This group of thinking cells (neurons) can generate a rhythmic pattern of motor activity without any feedback from the muscles having to traffic to the brain. Central pattern generators regulate movement like swallowing, breathing, and locomotion. What makes central pattern generators so amazing is scientists are trying to utilize them to generate movement in people with spinal cord injuries.

Scientists are researching how to activate central pattern generators in people with spinal cord injuries or in people where walking is difficult due to problems with the thinking cells (neurons). This research can result in some amazing ways to use central pattern generators to restore some ability to walk.

3.2 BACKGROUND INFORMATION FOR THE ACTION POTENTIAL

We know which regions of the brain are utilized in controlling movement, including force, direction, and timing. But how do those regions of the brain communicate with each other and communicate with the muscle cells? Thinking cells (neurons) use the action potential as a way to communicate with other regions of the brain and with other cells that they connect with to generate behavior. The action potential is what the thinking cells (neurons) use to activate muscles and generate motion.

We have approximately 100 billion brain cells including thinking cells (neurons) and supporting cells (glia). Thanks to work done by Camillo Golgi and Santiago Ramón y Cajal, it is known that those brain cells are separate from each other. Because they are separate from each other and not just one big mesh, they have to have a way of communicating with each other and getting all the coordinating brain regions on the same page. The structure of the thinking cell (neuron) is best suited for its function—communicating with its neighbors including muscles. The sending structures (axon) and receiving structures (dendrite) are used by the thinking cell (neuron) to communicate with neighboring thinking cells (neurons), which keeps the brain functioning as a unit.

3.2.1 Up to 100 Inputs

A single thinking cell (neuron) can receive information from up to 100 other cells.

Your brain is doing that right this minute at lightning fast speeds. Your eyes are taking in the contrast between the page and the text, communicating light and dark places, which are interpreted into letters, then into words, and those words represent some sort of meaning. Parts of your brain are determining what the words on the page are, other parts are determining the words overall meaning, and parts of your brain are working through understanding the meaning of the words. And still other parts of your brain are absorbing information about your surroundings, the sounds you are hearing, if you are hungry or tired, if you are sitting or standing, etc. The brain is amazing at how much information it is taking in and working with all at one time. In order to communicate all that information, thinking cells (neurons) use action potentials.

3.3 CHEMISTRY AND UNBALANCED IONS

How does the thinking cell (neuron) generate an action potential necessary to generate movement?

3.3.1 Electricity due to Imbalanced Ions

Elements have a charge, and so they are also called ions. Ions are molecules with a charge. In cells, there are many different ions each with very important jobs—sodium (Na+), potassium (K+), chloride (Cl−), and calcium (Ca2 + or Ca + +) are of importance to understanding the action potential.

Ions favor being in balance (equilibrium)—same number (concentration) of a particular ion inside and outside a cell (chemical concentration equilibrium) *and the* same charge inside and outside a cell (electrical equilibrium).

The thinking cell (neuron) has the ion concentration and charges out of balance. There is more sodium (Na+) outside the thinking cell (neuron) and more potassium (K+) inside the thinking cell (neuron). Throughout the rest of the body, ions are kept in balance (equal concentrations) inside and outside. Ions favor things being balanced.

Why does the brain intentionally keep ions out of balance? The thinking cell (neuron) can respond to a stimulus quickly by moving from "out of balance" (which is *not* the way ions like) to a "balanced" situation (which is what the ions would prefer). By keeping things "uncomfortable," the cell can react quickly because it wants to get to a "comfy" place. Think of it this way. If you are at the beach in the summer and you are standing barefoot on blazing hot sand, you run to the ocean at the speed of light so that your feet can cool off. If you had shoes on, you would not run like a crazy person to the ocean. You would take your time. Your thinking cells are barefoot on hot sand and want to move to the ocean. This is ideal for thinking cells (neurons) because it means they respond quickly. We really do not need our thinking cells (neurons) to meander or be sluggish in responding to a stimulus. And depending on if the ions move (either into or out of the cell), the thinking cell (neuron) knows what to do.

The membrane that surrounds the thinking cell (neuron) has electricity (membrane potential). That electricity of the membrane (membrane potential) is in the membrane of the thinking cell (neuron) because of the electrical charges of the ions found both inside and outside the cell (electrical gradient). Because of the ions being out of balanced, the thinking cell (neuron) has a more negative charge inside the cell than outside the cell.

The electricity (membrane potential) found in the membrane is necessary for the action potential, which is the main ways the thinking cell (neuron) communicates. The action potential is basically a big shift (from negative charges to positive charges) in the electricity (membrane potential) found in the membrane of the thinking cell (neuron), which makes the thinking cell (neuron) more positively charged. Before the action potential officially makes the big shift in the electricity of the membrane (membrane potential), there is a tiny shift in the electrical charge, which opens channels in the membrane of the thinking cell (neuron) that allow the ions to move, which causes the electricity of the membrane (membrane potential) to build a big change in electrical output, thus generating an action potential. If for some reason, the ions do not move or do not move enough and there is not a big change in the electricity in the membrane (membrane potential), then an action potential does not happen. There is no such thing as a half action potential. Action potentials are all or nothing, and nothing in between.

3.3.2 In and Out

In order to have an action potential, ions have to move in order to achieve balance. Where do these ions move? Through tiny gates, called ion channels, set up in the membrane of the thinking cell (neuron). These ion channels allow specific ions to either enter or exit the cell in order to achieve balance of that ion, and by doing so allow the cell to know how to respond to the stimulus. They are specific to that one type of ion to protect the thinking cell (neuron). Because they only let that one type of ion move either in or out of the cell, these channels do not let anything else in or out of the cell, which protects the cell from getting rid of something unnecessarily or letting in harmful materials.

To date, over 100 types of ion channels have been discovered. Most of the time these channels remain closed, like a door. Like a door, they have a lock. There are a few different doors in the membrane, each with its

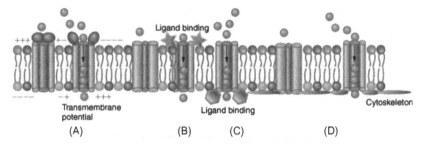

Figure 3.4 *Classes of ion channels stimulated by different gating mechanisms.* Ion channels are distinguished according to the signal that opens them. (A) Voltage-gated channels require a deviation of the transmembrane potential. Ligand-gated receptors respond to the binding of a specific ligand, either (B) an external neurotransmitter molecule or (C) an internal mediator such as a nucleotide or ion. (D) Mechanically gated channels can sense movement of the cell membrane linked by cytoskeletal filaments to the channel protein. Each effector causes an allosteric change that opens the channel, thereby causing an ion flux across the membrane.

own lock. Each lock has its own key. Some channels open when the electrical charge (voltage) of the membrane changes—these types of channels or pores are called voltage-gated channels or voltage-sensitive channels (Fig. 3.4A). Some channels open with those tiny spheres we see on the anti-depressant commercials fit into the lock on the channel. Those spheres are a chemical message (neurotransmitter), called a ligand. So these types of channels are called ligand-gated channels (in the image above, the neurotransmitter is actually a star (Fig. 3.4B)). The electrical or chemical signal then opens the channel and ions move so that they have the same concentration inside the cell as outside the cell.

The open channel causes a shift in one or more of the ions we mentioned (sodium, potassium, calcium, or chloride) and based on that shift in ions, the thinking cell (neuron) knows what to do. The bottom line is that there is a specific signal that opens the channel and the open channel moves the ions and the movement of ions results in a message for the thinking cell (neuron).

3.3.3 Recap

The action potential is a shift of the electricity found in the membrane of the thinking cells (neurons). This shift in electricity is due to the shift of ions—sodium flooding the cell and potassium escaping the cell through the channels. And these ions move because they favor being in balance the concentration of sodium and potassium. These ions also favor the cell to have an equal balance of electrical charges. This shift in electricity is how the thinking cells (neurons) communicate with neighboring cells and how they communicate the information about movement with our muscles.

3.4 HISTORY OF THE ACTION POTENTIAL: GIANT SQUIDS!

How did scientists figure out the action potential was necessary to communicate messages? There are several tightly controlled experiments that were done to demonstrate that ions moving created a change in the electricity of the membrane (membrane potential) and that the change in electricity (action potential) was the key to the thinking cells (neurons) communication.

Hodgkin and Huxley performed the most iconic experiments in the 1940s based on Ohm's law of permeability. They utilized a squid, not because those are readily available, but because the squid has a massive thinking cell (neuron) that is easily accessible in a dissection. Using the squid and some electrical equipment, Hodgkin and Huxley measured the flow of ions in and out of the squid's thinking cell (neuron). In their initial measurements, they were looking at the overall flow of ions—the electrical flow. But they had no idea which ion was going which direction. To figure out if sodium or potassium was leaving the cell or entering the cell during an action potential, they used different chemicals to block some of the ion's ability to flow. There are chemicals that block the channels for sodium and there are separate chemicals that block the channels for potassium. Initially, they blocked just the sodium channels (pores) and observed what happened to the electrical charge. Then they blocked the potassium channels (pores) and observed what happened to the electrical charge flow. They compared these results to what happened without any chemicals in the thinking cell (neuron) of the squid. Based on these experiments, they determined which ions had to flow in (sodium) or out (potassium) and when in order for a thinking cell (neuron) to generate an action potential.

3.5 CHEMICAL MESSAGES ASSOCIATED WITH EXERCISE

The brain uses the action potential to generate the communication necessary to generate movement, like movement in exercise. That is *how* the brain generates movement. What does that movement do for the brain? Thanks to *Legally Blonde* (2001), we are aware that "Exercise gives you endorphins. Endorphins make you happy." Endorphins are part of the chemical messages (neurotransmitter) in your brain. During exercise, the brain releases endorphins, which act as a sedative and reduce the feeling of pain, both of which reduce the feeling of stress and anxiety. In this way, exercise can shut down a stress response. A lot of runners refer to the release of endorphins as a "runner's high" because of the release of the stress during exercise.

How powerful are those endorphins? Pain killers, nicotine, and abused substances structurally mimic the chemical messages that occur naturally in the brain. Morphine mimics those exercise-released endorphins. The difference between endorphins and morphine is that morphine will also activate other pathways, which result in addiction, whereas exercise-induced endorphin release is not addictive. The behavior of exercise can be addictive, but that is not due to the endorphin release. The bottom line is that endorphins are powerful sedatives and provide a natural way to reduce stress and allow the brain to relax.

Not only does the endorphin release reduced anxiety and stress, resting the brain while we focus on the activity stops the cascade of anxiety and stress. By focusing on something other than what is on the "to do" list, we give our brain a necessary and very healthy pause. This pause stops the brain's reaction to stress and anxiety. Usually when we exercise, we are not interrupted by the issues we are facing. So many forms of exercise afford us the ability to focus on the exercise itself or the environment we are in.

3.6 WHAT HAPPENS WHEN THE MESSAGE FOR MOVEMENT ISN'T COMMUNICATED?

The action potential is the way that thinking cell (neuron) communicates in particular, it generates the movement necessary for exercise. The action potential is necessary for the brain to function properly. Without proper action potentials, several disorders result such as multiple sclerosis (MS) and seizures.

3.6.1 Multiple Sclerosis

Multiple Sclerosis (MS—letters said indvidually) is an inflammatory disease that affects the central nervous system (CNS). It is thought that there is a genetic and environmental cause of the disease, but at the moment, there is no known cause for MS. This disease usually affects people around 20—40 years old, with more women than men being diagnosed. The body mounts an immune response attack on itself, on the CNS. The sending structure (axon) has a special wrapping made of the supporting cell (glia) oligodendrocytes (called myelin) around it. That wrapping insulates the sending structure (axon) and makes the action potential conduct faster and without interruption. In MS, the body attacks that wrapping (myelin), forming a scar (sclerosis) on the thinking cell (neuron), making it harder and harder for an axon to send the electrical message correctly.

Because MS attacks how the cells communicate, there are a wide variety of symptoms which are determined by whatever region of the brain loosing myelin. Some common symptoms of MS are sensory loss, weakness, pain, double vision, imbalance, vertigo, and bladder dysfunction. Most people with MS retain the ability to walk, usually assisted. With advances in medicine and therapies, the life span of people with MS is approximately 7 years less than persons without MS.

There are four patterns of MS progression: relapse-remitting MS, primary progressive MS, secondary progressive MS, and primary relapsing MS. The patterns are based on when the symptoms flare up (relapses) and when the symptoms are in remission, and these periods are very unpredictable for most people. Treatment tries to minimize the relapse and extend the remission, but there is currently no cure. Most of medications are aimed at reducing the inflammation in the CNS. Because MS is an immune response, the brain becomes inflamed, which is not a healthy state of being for the brain. Other medications aim at managing the side effects of the disorder. And now exercise is being considered as part of a treatment plan. Initially, it was thought exercise would increase side effects of disease. But more and more research is demonstrating the benefits of exercise, especially yoga and Pilates, for people with MS.

Annette Funicello was a Disney starlet who was diagnosed with MS. There is a movie documenting her fight with MS *A Dream is a Wish Your Heart Makes*. Another documentary *Hiliary and Jackie* describes the battle of classic cellist Jacqueline du Pre with MS.

3.6.2 Seizures

The first account of seizures date back to the late 1800s when it was thought that seizures were the manipulation of the person by a spiritual power. When a thinking cell (neuron) doesn't send an action potential on time or it is the action potential is altered in strength (any form of abnormal activity), it disrupts that region of the brain. If the abnormal activity is synchronized between enough cells, a seizure can result either in a specific brain region or in the entire brain. There are three common symptoms of a seizure—a warning of some sort (odor or sensation or feeling), a loss of consciousness, and a motor component (shaking, hand rubbing, and/or chewing).

Seizures can be grouped based on if there is a specific cause (symptomatic seizure) or if there is no cause (idiopathic seizure). There are several different types of seizures: focal seizures, generalized seizures, absence, and tonic–clonic seizures. Focal seizures originate from a relatively distinct part of the brain and always involve one hemisphere. Generalized seizures involve both hemispheres and involve networks of neurons. Absence seizures are repeated periods of absence or daydreaming. Tonic–clonic seizures are spontaneous and defined by tonic muscle contraction.

Seizures can be their own disorder or they can be a symptom of other neurological disorders. One seizure disorder is epilepsy. Epilepsy is spontaneous, repeated, and unprovoked (idiopathic) seizures. Usually, epilepsy is diagnosed after three or more seizures. Another example of a seizure disorder is Dravet syndrome. People diagnosed with Dravet syndrome have a genetic mutation in a sodium channel, altering how the thinking cells (neurons) communicate, resulting in seizures. Currently, treatments for Dravet syndrome have made the news because medical marijuana reduces the number of seizures from the hundreds and thousands that might occur in a month to single digits. Obviously, increased work understanding the long-term use of medical marijuana in this patient population is necessary, but it is evident that decreased number of seizures can greatly impact the people ability to function at a higher level.

There are several different documentaries on living with seizure disorders, including *On the Edge: Living with Epilepsy* (2016).

3.6.3 What Happens When the Brain Cannot Control Movement?

Parkinson's disease (PD-letters said individually) is a neurodegenerative movement disorder. Thinking cells (neurons) in a specific region of the brain (substania nigra) begin dying, but there is no known cause for the death of these cells. These cells send out a chemical message (dopamine) that controls muscle movement.

Because there is a loss of the dopamine within the substance nigria, people with PD experience tremor in their hands, arms, legs, and/or face, stiffness in the limbs and/or trunk, slowness of movement and postural instability or impaired balance and coordination. To get a better sense of what this disease looks like, there are many documentaries listed on the World PD coalition website.

PD is usually diagnosed in people aged 55–60. The symptoms are subtle and appear gradually. The disease progresses at different rates in different people. There is no known cause or cure for PD. It is a chronic disorder. Some people with PD become disabled, while others have minor complications due to the disease. Currently, it is unknown which symptoms will impact a patient and at what level of severity.

There are medications that relieve some of the symptoms of PD. In particular, levodopa (L-DOPA-the letter L is said individually, then do-pa is said as a word.), which is the precursor for dopamine. Because this disorder is a result of the loss of the dopamine-containing thinking cells (neurons) in the substania nigra, it stands to reason that replacing that dopamine can alleviate some of the symptoms. Dopamine cannot cross the blood brain barrier (BBB), but it's precursor, L-DOPA, can. Eventually, too many cells in the substania nigra are lost and L-DOPA treatment becomes less effective. Another treatment option in some cases is the implantation of a deep brain stimulator in the substania nigra. The stimulation activates what cells are left and can help alleviate the loss of signaling that results from loss of cells in the substania nigra.

3.7 CONCLUSION

The brain uses multiple regions to generate direction, force, timing, and execution of movement. Signals sent between thinking cells (neurons) and other thinking cells (neurons) or the muscles are action potentials, which are changes in electricity due to movement of ions across the thinking cell (neuron)'s membrane. There are several movement disorders that occur when these action potentials are dysregulated, including MS, seizure disorders, and PD.

Why should exercise and movement be a part of our every day? Exercise not only helps muscles, the profound calming effect it can have on the brain (and thus ultimately the heart) should make it a top priority for a portion of each day. If we want to improve at school or excel at work, a 20-minute break to exercise and relax the brain can improve learning and memory.

BIBLIOGRAPHY

Augustine, J.R., 2007. *Human Neuroanatomy*. Elsevier Academic Press.

Baars, B., Gage, N., 2010. Cognition, Brain and Consciousness: An Introduction to Neuroscience, second ed Elsevier Academic Press.

Goodman, S.R., 2007. Medical Cell Biology, third ed Elsevier Academic Press.

Kandal, E.R., Schwartz, J.H., Jessel, T.M., 2000. Principles of Neural Science, fourth ed McGraw-Hill Companies.

Kolb, B., Whishaw, I.Q., 2014. An Introduction to Brain and Behavior, fourth ed Worth Publishing.

Mason, P., 2011. Medical Neurobiology. Oxford University Press.

Nicholls, J.G., Martin, A.R., Fuchs, P.A., Brown, D.A., Diamond, M.E., Weisblat, D., 2012. From Neuron to Brain, fifth ed Sinauer Associates, Incorporated.

Purves, D., 2012. Neuroscience, fifth ed Sinauer Associates, Incorporated.

Sontheimer, H., 2015. Diseases of the Nervous System. Elsevier Academic Press.

Squire, L., Berg, D., Bloom, F., du Lac, S., Ghosh, A., Spitzer, N., 2012. Fundamental Neuroscience, fourth ed Elsevier Academic Press.

Squire, L., Kandel, E., 2009. Memory: From Mind to Molecules, second ed Roberts and Company Publishers.

CHAPTER 4

9:00 a.m. Lions and Tigers and Bears, Oh My! Oh Wait, No, It's Just Work!

Contents

SUMMARY

Stress is an everyday issue. Like pain, stress is necessary to keep us out of danger. However, too much stress can cause problems. How does the brain differentiate between good stress and bad stress? The chemical messages (neurotransmitters) of the brain indicate the difference between the good stress and the bad stress. When an action potential is generated, that action potential releases chemical messages (neurotransmitters) to tell neighboring cells different information. There are different types of chemical messages (neurotransmitters) and those messages have to have a receptor to interact with. What happens when the chemical messages (neurotransmitters) are messed up? Disorders such as general depression, bipolar disorder, and chronic stress can result.

Neuroscience Basics.
DOI: http://dx.doi.org/10.1016/B978-0-12-811016-4.00004-0

4.1 STRESS

4.1.1 Normal Stress

A stressor is defined as a stimulus that challenges our brain and body out of baseline operations. It's anything that causes the brain or body to shift to an out of normal range. Changes out of the normal range are regulated by chemical messages (neurotransmitters) in the brain. The brain uses chemical messages to communicate information about the internal and external environment. Stress is no different. There are two phases or types of a stress response—a fast response driven by the chemical message (neurotransmitter) epinephrine and a slow response driven by the chemical message (neurotransmitter) cortisol.

The fast response is governed by the sympathetic nervous system, often referred to as the flight or fight response. In a situation where we sense danger, our brain very quickly assesses if you are going to fight the danger or flee it. The sympathetic system prepares the body for intense physical activity by elevating heart rate, stimulating sweat glands, inhibiting digestion and other responses. An example—the hypothalamus can send a message that a flight or fight response is necessary. This stimulates the medulla, which releases epinephrine into the system, which will stimulate the other glands of the sympathetic pathway.

The slow response to a stimulus is regulated by the parasympathetic system, also known as "rest and digest." The body focuses on the stimulus so it can address and solve the problem as well as repair any damage caused by the stress. It relaxes the body and slows down high energy functions so the body can rest and recover. It maintains the levels of "normal" for all the main systems of the body. An example here is a release of cortisol into the blood stream, which slows the heartbeat, stimulates digestion and other organs of the endocrine system.

Both responses are generated from a region of the brain (hypothalamus), which produces hormones. And, by way of a reminder, in chapter 1, we discussed that the hypothalamus is part of the diencephalon. The hypothalamus regulates eating, drinking, growth and maternal behaviors. Based on the amount of the hormone and which hormone is generated in the hypothalamus, the brain can determine if the stress response needs to be fast or slow. The hormones act as a chemical message (neurotransmitter), they open those channels in the membrane that tell the thinking cells (neurons) what to do. In order to better understand the stress response, we first need to understand how chemical messages regulate the thinking cells (neurons).

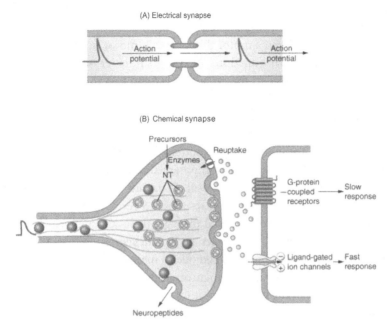

Figure 4.1 *Electrical and chemical synapses.* In an electrical synapse (connection) (Fig. 4.1A), a gap junction which physically connects the two thinking cells (neurons) allows current to flow directly from the sending cell (presynaptic cell) to the receiving cell (postsynaptic cell). In a chemical synapse (Fig. 4.1B), the thinking cells (neurons) are not physically touching. There is a space between them (the synapse or synaptic cleft). In a chemical synapse, the sending cell (presynaptic cell) converts the current into a chemical signal which is sent to the receiving cell (postsynaptic cell). Neurotransmitters (NT) (light gray), located in synaptic vesicles, and larger neuropeptides (dark gray), located in dense core granules, are the chemical signal. The chemical messages are released into the synaptic cleft and interact with receptors on the receiving cell (postsynaptic cell).

4.2 BACKGROUND INFORMATION ABOUT NEUROTRANSMITTERS

How does the brain "know" it is stressed? How do chemical messages (neurotransmitters) communicate flight or fight or rest and digest? Most thinking cells (neurons) use chemical messages to talk to their neighbors, through a small space between the neighbors called a synapse. What is a synapse? When one thinking cell (neuron)'s axon communicates with the receiving structure (dendrite) of a neighbor cell, they form a synapse. The synapse is made up of the axon (sending structure), the dendrite (receiving structure), and a space between those two called the synaptic cleft. A chemical synapse is set up to respond to chemical messages (neurotransmitters). What this means is that several receptors are on the receiving structure (dendrite) of the synapse ready to respond to chemical messages (neurotransmitters) that get released. Synapses between a motor thinking cell (motor neuron) and a muscle are a neuromuscular junction (NMJ—each letter said individually). These synapses are similar to the synapses between two thinking cells (neurons). At the NMJ, the chemical message (neurotransmitter) is released and there are receptors on the muscle that the chemical message (neurotransmitter) binds to and tells the muscle what to do.

The chemical messages (neurotansmitters) in the thinking cells (neurons) can use the electricity from the action potential to release chemicals (neurotransmitters) which communicate as well. What are these chemical messages in the brain that determine the brain response to certain stimulus?

There are several types of chemical messages (neurotransmitters). Chemical messages (neurotransmitter) have rules and guidelines about what makes a chemical message (neurotransmitter) and official messenger in the brain.

Criteria of chemical messages (neurotransmitters):

1. Must be present in the sending structure (axon) of a thinking cell (neuron)
2. Must be released from a thinking cell (neuron) in response to an action potential
3. Must bind and activate a receptor (special kind of channel).

4.3 TYPES OF CHEMICAL MESSAGES (NEUROTRANSMITTERS)

To keep things simple, we are going to focus on a few of the main chemical messages (neurotransmitters). The first two we are going to discuss are the ying and yang of the brain—GABA (pronounced Ga-Ba) and glutamate. GABA is a chemical message that communicates "stop." You can think about GABA as the brakes. Glutamate is a chemical message that communicates "go." You can think about glutamate as the accelerator. Both GABA and glutamate are required in proper amounts for proper function.

Common Neurotransmitters

GABA—primary inhibitory
Glutamate—primary excitatory
Acetylcholine—memory, heart rate, muscle and motor control
Norepinephrine—attention, alertness, circadian rhythms, memory, moods
Serotonin—mood and circadian rhythms
Dopamine—motor control and reward.

Another chemical message (neurotransmitters) is acetylcholine, which regulates memory storage, slows down the heart rate, activates muscle contraction, and regulates motor control. Norepinephrine regulates attention, alertness, circadian rhythms, memory formation and moods. Serotonin regulates mood and circadian rhythms. Finally, dopamine regulates motor control and reward. Each of these chemical messages (neurotransmitters) has a variety of functions specific for a particular brain region. For instance, we will talk about the reward pathway and addiction. In that case, the main chemical message (neurotransmitter) responsible for regulating normal function of the reward pathway is dopamine. Dopamine is also the main chemical message (neurotransmitter) responsible for regulating another region, the basal ganglia, which regulates parts of motor control.

4.4 LOCK AND KEY

Each chemical message (neurotransmitter) has its own receptor. The chemical message (neurotransmitter) is the key and the receptor is the lock. In the previous chapter, we mentioned that there are ligand-gated channels and voltage-gated channels. The chemical message (neurotransmitter) binds to the ligand-gated channels. When it binds, a channel (receptor) opens and results in a specific action on the part of the thinking cell (neuron).

So why are receptors important? Exciting as they are, especially the voltage-gated ones (pun intended), several different drugs and poisons target the channels. If channels are manipulated, it changes how the brain works.

Here are some examples of how channels (receptors) for chemical messages can be altered. Local anesthetics work by blocking sodium channels that would usually communicate pain. This is a good thing when undergoing surgery or dental procedures. Tetrodotoxin, found in puffer fish, blocks voltage-gated sodium channels. Puffer fish (in Japanese, Fugu) is a delicacy in some parts of the world and if not prepared properly, the food can contain dangerous levels of the poison. The practice of eating fugu in Japan dates back over 2000 years ago and as early as 4000 BC in China. Fugu is the only food the emperor of Japan is forbidden to eat. Depending on how much tetrodotoxin is consumed, you can experience numbness and shortness of breath all due to the body's inability to generate action potentials because sodium channels are blocked. Approximately 20−40 people die of fugu poisoning each year in Japan and approximately 30−60 were hospitalized due to fugu poisoning. There is no antidote. Treatment for fugu poisoning is mechanically breathing for the affected individual until the poison has left the body and the brain can resume the task of controlling the lungs. Another side note about tetrodotoxin. Remember back to the experiments Hodgkin and Huxley did on the squid giant axon to determine which direction the ions were flowing? They blocked the ability of sodium to flow and they blocked the ability of potassium to flow. Well, they blocked the sodium flow using tetrodotoxin. In fact, scientists today use tetrodotoxin in experiments to examine sodium ions flowing in different conditions.

4.5 STOPPING THE CHEMICAL MESSAGES

Once the chemical signal is released, the signal must be communicated to the receptor and then the signal must be shut down. You don't want the message "stop" to turn into "stop, stop, stop, stop, stop." Those are two different messages. So how do we shut down the chemical message? How do we turn it off? There are two main ways—pores and chemicals.

There are pores that usually take the chemical message (neurotransmitter) back up into the sending cell and then the chemical message (neurotransmitter) gets recycled within the sending cell. The other way to shut down the chemical message is to chemically break it down in the space between the sending and receiving cell. As with other parts of sending chemical messages (neurotransmission), we can alter the amount of message being received by a thinking cell (neuron) by altering the pores or the chemicals breaking down the chemical message (neurotransmitter). Some drugs work by eliminating or increasing the chemicals necessary to break down the chemical message (neurotransmitter). Other drugs can work by blocking the pore on the sending cell, increasing the amount of chemical message (neurotransmitter) present. An example of the latter: serotonin reuptake inhibitors block the serotonin pores on the sending cell increasing the amount of serotonin present.

4.6 TYPES OF COMMUNICATIONS

Thinking cells (neurons) use chemical synapses to send signals. The number of receptors can impact that message as well as the amount of chemical messages that are released. Another way to impact the chemical signal is through signaling pathways.

4.6.1 Microphones and Amplifiers

Thinking cells (neurons) can utilize another way of communication in order to better control the timing of the message or the amplification of the message or both. Thinking cells (neurons) have a set of proteins (secondary messenger signaling cascades) whose job is to pass of the message to other parts of the cell. Protein #1 receives the message and tells proteins #2 and #3. Protein #2 tells proteins #4, #5, and #6. Protein #3 tells proteins #7, #8, and #9. And so on and so forth. These proteins are set up to communicate a message or a signal that regulates how the thinking cell (neuron) functions. And this message regulates the timing and the amplification of the message. In another scenario, Protein #1 may only tell protein #2, so Proteins #7, #8, and #9 don't get the message. These messaging cascades are tightly controlled and are another way that thinking cell (neuron) receives the message from a neighboring cell. And these cascades are set off when a chemical message (neurotransmitter) binds to a special receptor that is not a channel (metabotropic receptor) (Fig. 4.2).

We know that the body regulates functions like stress with chemical messages (neurotransmitters). And depending on which message is

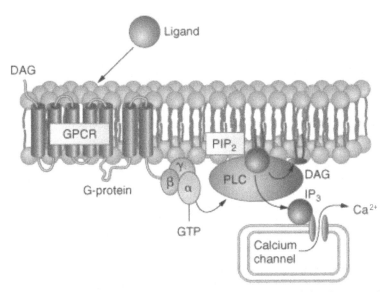

Figure 4.2 *G-protein–mediated signal transduction.* A ligand binds to a G-protein–coupled receptor (GPCR) in the membrane. This, in turn, activates a phospholipase C, which hydrolyzes phosphatidylinositol biphosphate (PIP$_2$) to form diacylglycerol (DAG) and inositol triphosphate (IP$_3$). IP$_3$ acts to increase cytosolic calcium as part of a signal transduction cascade.

released, a short or long term stress response is elicited. We know what chemical messages (neurotransmitters) are and that they are released into a synapse after the action potential. They travel across the synapse to the receiving cell and bind to a receptor. That receptor may be a channel or it may be a repeal receptor that sets up a secondary messenger signaling cascade. Either way, the chemical message (neurotransmitter) elicits a response from the neighboring cell.

4.7 WHAT HAPPENS WHEN CHEMICAL MESSAGES ARE MESSED UP?

There are several disorders that are a result of a chemical message (neurotransmitter) failing to send properly, too much or too little of a chemical message (neurotransmitter) present in the brain or a brain region. Two of those disorders are depression and bipolar disorder (also known as bipolar depression or manic depression).

4.7.1 Depression

Thanks to pharmaceutical commercials and the prevalence of depression in our society, general depression is fairly familiar to most of us. Depression is seen as early as the teenage years, but most prevalent in middle-aged population. Depression causes severe symptoms which impact how a patient feels, thinks, or able to balance and commit to daily life tasks. It is characterized by prolonged feelings of worthlessness and guilt, loss of interest in hobbies, persistent sad or anxious thoughts, irritability, loss of focus, changes in memory, aches and pains or headaches without a clear physical cause. These symptoms can result in a disruption of normal eating habits, sleep habits and a general slowing of behavior. Taken together, these symptoms impact quality of life. There is no known cause, either genetic or environmental. There are many documentaries depicting the symptoms of a person with depression. One of the documentaries is Depression: Out of the Shadows (2008).

There are chemical messengers (neurotransmitters) that impact the disease (norepinephrine and serotonin) but research has not yet demonstrated that alterations those chemical messengers (neurotransmitters) cause the disease. We know norepinephrine and serotonin impact the disease because several of the pharmaceuticals that target depression aim at elevating norepinephrine or serotonin levels. Drugs can do this a few different ways. The drug can block whatever clears the chemical message (neurotransmitter) from the synapse (a pore, or the chemical that breaks down the chemical message) or a drug can block any signal that might inhibit the release of norepinephrine and/or serotonin.

We talked about two ways chemical messages (neurotransmitters) are cleared from the synapse. The first was taking the chemical message (neurotransmitter) back up through a channel on the sending thinking cell (neuron) in the synapse. There is an entire class of depression medications that block the reuptake of chemical messages (neurotransmitters) into the sending cell—selective serotonin reuptake inhibitors or SSRI (the letters are pronounced individually for this acronym) and serotonin norepinephrine reuptake inhibitors or NSRI (the letters are pronounced individually for this acronym). There is another class of antidepressants called tricyclic antidepressants (TCA—the letters are pronounced individually) that blocks one chemical message (acetylcholine) to increase the chemical message (neurotransmitter) for norepinephrine and serotonin.

The other way chemical messages (neurotransmitters) can be broken down in a synapse is through chemical means. There are chemicals designed to break apart a chemical message (neurotransmitter) so that it does not keep sending a signal in a synapse. There is an entire class of antidepressants that block the activity of the chemical (Monoamine oxidase) that breaks down norepinephrine and serotonin (and dopamine). This class of antidepressants is called monoamine oxidase inhibitors (MAOIs—the letters are pronounced individually).

4.7.2 Bipolar Disorder (aka Bipolar Depression or Manic Depression)

Two documentaries that demonstrate the life of a person with bipolar disorder are *Stephen Fry: The Secret life of the Manic Depressive* (2006) and *Biography: Patty Duke* (2003). Bipolar disorder has extreme mood swings that alternate between euphoria (mania) and depression, more so than usual ups and downs in life. The cause of bipolar disorder is not known, however, there is some evidence that it may be inherited. Events such as stress, substance abuse, and sleep deprivation may contribute to the cause of the disease. The first episodes of bipolar disorder are usually observed in the teenage years of early adulthood.

Euphoria (mania) is associated with an over-inflated self-esteem and grandiosity. This will eventually lead to paranoia and delusion. During the manic state, people are hyperactive, enthusiastic, full of high expectations and yet they fail to follow through with most of their expectations. They feel invincible and destined for greatness. Eventually, this will spiral out of control into recklessness, irritability, or poor choices. Depression in bipolar disorder shares symptoms with general depression. But there are certain symptoms that are more common such as irritability, guilt, and restlessness.

The cycles between mania and depression can last for days, weeks, or months. The changes in the moods are so intense that the interfere with normal everyday functioning. The cycles and severity of the moods vary from person to person. Treatment is a combination of therapy and medication to manage the manic and depressive phases of the disorder that alter the levels of chemical messages (neurotransmitters) in the brain.

4.7.3 Stress Gone Wrong

How does the brain usually shut down a stress response? Usually the learning and memory region of the brain (hippocampus) is in part responsible for shutting down a slow stress response to protect the brain. The learning and memory region (hippocampus) detects the level of the stress hormones and tells the brain region producing the stress hormone (hypothalamus) to stop making the hormone. Chronic stress arises when the stress hormone is still made, potentially causing lasting damage to the region of learning and memory (hippocampus) (Fig. 4.3).

Many Americans are plagued by chronic stress—stress that just won't go away. Reports vary greatly on how many Americans are living with chronic stress, but consistently, chronic stress is on the rise. Chronic stress starts out with an initial stressor that would usually generate a slow stress response, but then the stress response won't turn off due to poor

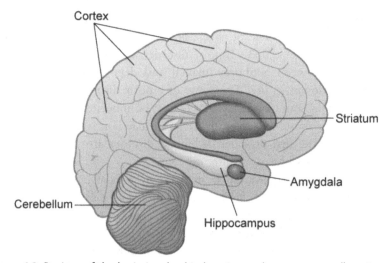

Figure 4.3 Regions of the brain involved in learning and memory as well as stress.

nutrition, inadequate sleep, or continued emotional distress leading to consistently high cortisol levels. So if you have an initial stressor like a medical problem or a bill or a school/work project and then you couple that with poor nutrition and inadequate sleep, chronic stress can develop.

Why is a prolonged slow stress response, or chronic stress, a bad thing for the brain? Chronic stress decreases our ability to learn or recall information. This is because the hormones that the brain generates due to stress events (cortisol) change the area of the brain (hippocampus) that is related to learning and memory, which results in a decrease in the ability to learn or recall information. So chronic stress decreases learning and recall.

Chronic stress decreases our ability to pay attention. This can be due to several changes in the brain. Decreased attention could be due to focus on the stressor or loss of sleep. It may also be due to changes in the brain caused by increased cortisol. Whatever the reason, chronic stress not only decreases learning and recall, it also decreases attentiveness.

Chronic stress will decrease a proper perception of reality. This could be due to increased cortisol levels in the brain or due to the loss of sleep or both. Altered reality due to chronic stress can be presented in several ways: we may feel like something isn't right, that we are lost or confused, that we cannot process information, or that we are watching ourselves from outside our body. Together, chronic stress results in an altered perception of reality as well as deceased attentiveness, learning and recall.

Chronic stress can also result in moodiness, irritability, depression, aches, nausea, frequent colds, insomnia, muscle pain, high blood pressure, weakened immune system, heart disease, digestive problems, and obesity. The bottom line is that the side effects of chronic stress are negative and can feed into an increase in a stress response as a person becomes stressed out about the side effects of chronic stress as well.

4.8 CONCLUSION

We have examined why normal stress is healthy, and how the brain uses chemical messages (neurotransmitters) to determine how to respond to different stimuli in the environment. While most of us have experienced depression first hand or through a friend or family member, statistically, even more of us have experienced persistent stress.

How can we reduce stress, and in particular, chronic stress? We see the negative side effects and don't want them! For most of us, there isn't much we can do to change the volatile economy, bills, deadlines, government, administrative red tape at work, stress of aging parents or raising kids or being alone. *So how do we battle the things that cause our chronic stress if we cannot reduce or eliminate the thing causing the stress?* Sleep, exercise, and eating right. It sounds trite and cliche, but hear me out. Sleep improves our ability to learn, remember, and be alert. Sleep can help combat the side effects and some of the cause of chronic stress. When we exercise, we are focused on what we are doing, which gives our brain a break from the stress. Exercise also releases endorphins which act as a mild sedative, calming the brain. Finally, eating right gives our brain the proper nutrition necessary for peak performance. When we focus on our health, we are not focusing on the cause of our stress. This gives our brain a break and helps reduce some of the stress. This is a small break in the cycle of chronic stress and can help combat chronic stress.

I know that one of the main causes of stress is lack of time, so the idea of adding exercise or going to bed earlier seems laughable. But if we do not have enough time to eat well, exercise, or sleep, than we are going to negatively impact our ability to learn, remember, be alert, and perceive the world properly. We all need to make time for sleep and exercise and eating right so that the chronic stress will not negatively impact our lives. These three keys of self-care can greatly reduce chronic stress and prevent permanent damage.

BIBLIOGRAPHY

Augustine, J.R., 2007. *Human Neuroanatomy.* Elsevier Academic Press.
Goodman, S.R., 2007. Medical Cell Biology, third ed. Elsevier Academic Press.
Kandal, E.R., Schwartz, J.H., Jessel, T.M., 2000. Principles of Neural Science, fourth ed. McGraw-Hill Companies.
Kolb, B., Whishaw, I.Q., 2014. An Introduction to Brain and Behavior, fourth ed. Worth Publishing.
Mason, P., 2011. Medical Neurobiology. Oxford University Press.
Nicholls, J.G., Martin, A.R., Fuchs, P.A., Brown, D.A., Diamond, M.E., Weisblat, D., 2012. From Neuron to Brain, fifth ed. Sinauer Associates, Incorporated.
Purves, D., 2012. Neuroscience, fifth ed. Sinauer Associates, Incorporated.
Squire, L., Berg, D., Bloom, F., du Lac, S., Ghosh, A., Spitzer, N., 2012. Fundamental Neuroscience, fourth ed. Elsevier Academic Press.

CHAPTER 5

10:00 a.m. Staff Meeting About the New Thing and How We Have to Learn It to Do Our Job!

Contents

SUMMARY

Throughout our lifetime, we are having to learn and remember information for a test or where we parked our car or a new way to do our jobs because of company changes. Whatever the reason, we are having to learn and remember as part of our everyday life. So what kinds of learning and memory are there? How does the brain learn? What has research taught us about learning and memory? How does aging impact our ability to learn and remember? And what disorders impact the ability to learn and remember?

5.1 TYPES OF LEARNING AND MEMORY

Remembering the past is a form of mental time travel: it frees us from the constraints of time and space and allows us to move freely along completely different dimensions.

Eric Kandel (father of Neuroscience)

Neuroscience Basics.
DOI: http://dx.doi.org/10.1016/B978-0-12-811016-4.00005-2

You have to begin to lose your memory, if only in bits and pieces, to realize that memory is what makes our lives. Life without memory is no life at all, just as an intelligence without the possibility of expression is not really an intelligence. Our memory is our coherence, our reason, our feeling, even our action. Without it, we are nothing.

Luis Buñuel

Before we start breaking down how we learn, let's go over some basic terms. Learning is a behavioral change due to an experience. Memory is storage and recall of information. Memory itself can be broken down into two types: explicit and implicit. Explicit (or declarative) memory is recall of personal events/personal facts (episodic memory) or recall of facts (semantic memory). Implicit (or nondeclarative) memory is recall of reflexive motor skills or perceptual skills (Figs. 5.1—5.3).

The saying goes "It's like riding a bike. Once you learn to do it, you will always know how to do it." For the most part, this saying is correct. Once we learn how to do a job or task or responsibility, we basically know how to do it for a lifetime. There are variables that change that impact learning but for the most part, once a task or responsibility is learned, it sticks with us.

Depending on your day-to-day job, you may use more of the explicit or the implicit memory. Some jobs that require skill sets: plumbers, surgeons, repair work, carpentry, etc. favor the implicit memory for the day-to-day motor tasks that are part of the job. Some jobs favor explicit memory such as those that require remembering facts and figures and assembling knowledge to solve a problem or teach a class. Humans

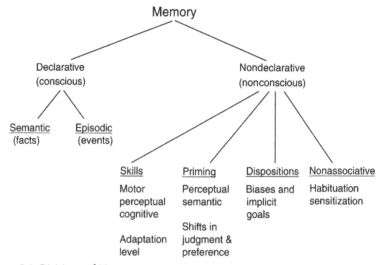

Figure 5.1 Divisions of Memory.

retrieve these memories with very little effort. And there are a lot of jobs that require both implicit and explicit memory.

The brain connections (synapses) due to activity (or inactivity) and stimulus, which results in learning. Scientists use a term to describe this— the brain is plastic or has plasticity. In this context, plastic or plasticity means the brain *can change*. In other words, no matter how old a dog is, you can teach it new tricks.

So learning is the structural change in brain connections (synapses) and memory is the storage and recall of past experiences.

A new memory is made in the amygdala or hippocampus. Again, by way of reminder, the hippocampus and the amygdala are part of the cerebral hemispheres. Both structures play a role in learning and memory. The amygdala also has roles in emotion and fear. The new memory formed in the amygdala or the hippocampus gets shipped off to a long-term memory storage in the cortex usually during REM sleep so long as there is sufficient nutrition to ship the memory from the short-term to the long-term location (Fig. 5.2). That region of the cortex will be specific for either implicit or explicit long-term memories. This is why sleep and proper nutrition are so important. In order to properly form a memory, you have to go through at least one sleep cycle and your brain has to have all the proper nutrition.

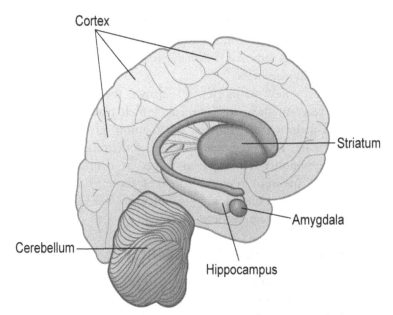

Figure 5.2 The Human Brain - showing areas necessary for memory - the hippocampus, the amygdala, and the cortex.

Learning is also reversible, which most of us are more than well aware of because we forget things. Forgetting can be just as important as remembering. There are a few things we actually need to forget. A very small percentage of people cannot forget. Some people have photographic memory, termed eidetic memory. The few cases of photographic memory are usually observed in children aged 1−7. Of children in that age group, 2−5% have reported photographic memory. If the ability is ignored, the photographic memory will fade. There are very few supported cases of photographic memory in adults.

5.2 MOVING THE CHANNELS

What happens during learning that can change the thinking cells (neurons) in the brain allowing us to remember? In a region of the brain that is necessary for forming new memories (the hippocampus and/or amygdala), certain chemicals open those channels and move the ions and by doing that, it makes those connections (synapses) within that brain region stronger. To make a thinking cell (neuron) stronger, the cell puts new channels into the membrane allowing it to respond to even more stimulus. A thinking cell (neuron) becomes stronger by increasing its ability to react. As a thinking cell (neuron) increases the strength of the connections (synapses), we form a new memory.

How do we know that it is those specific channels in the hippocampus? Scientists studied this in mice. In the region for learning and memory, the hippocampus, scientists removed the channels we discussed and the mouse was unable to create memories.

When we form new memories, we change the way thinking cells (neurons) can react to their environment, which makes them stronger. And to form a memory, the thinking cell (neuron) places new channels into the membrane. As you might imagine, forgetting or depressing a memory is the opposite. In order to forget or depress the memory, the thinking cell (neuron) removes some of those channels from the membrane, which makes the thinking cell (neuron) less reactive the environment, weakening the connections (synapses).

5.3 LEARNING AND MEMORY RESEARCH

Because learning and memory are a large part of the human experience, we are going to cover some of the research that is going on in the field of learning and memory.

With learning and memory, scientists are trying a variety of approaches to look at different aspects of learning and memory. The idea is that by looking at all these angles, scientists can understand the regions of the brain, the cells and the proteins necessary for learning and memory to occur properly. While we are focusing on learning and memory research, the various types of research from macro to micro are evident in just about every field of neuroscience research. We will also explore the organisms that scientists use for these studies, such as fruit flies, mice, and nonhuman primates.

A lot of experiments done with fruit flies produced a greater understanding of what is necessary for proper learning and memory. Fruit flies are a great system to use for studies, because their genetics are easy to manipulate and they have a short life cycle.

Mice are also used to study learning and memory. Mice genetics are not all that different from human genetics (let that one sink in for a moment). Mice and humans share approximately 85% of the protein-coding regions of their genes. Remember that all the cells in the body contain all the genes, but not all of those genes are turned on, or expressed.

Nonhuman primate research is less common but necessary at times because of the higher percentage of genes shared between humans and nonhuman primates. When scientists use animals several different regulatory boards governed by federal laws ensure that all conduct with the animals is ethical and necessary.

When looking at the types of learning and memory research published, the types of studies range from the macroscale to the microscale: observable behavior studies, whole brain studies, region-specific studies, cellular studies, and subcellular studies. Macroscale studies involve observing behavior or changes in brain regions or brain connectivity. The microscale is exactly that, studies which require microscopes or techniques to examine some of the small components of the brain.

On the macroscale end of research, one of the areas of learning and memory scientists study is behavior. Scientists may use human subjects and utilize brain scans during a behavioral test or use post-mortem tissue to compare regions that govern learning and memory. Scientists also use animal models in behavioral studies to research learning and memory. In mouse models, there are serval types of behavioral studies that examine different types of learning and memory. One of the most common types of behavioral studies with mice are mazes. In these mazes, mice have to learn where the reward or the safe place is. Then scientists can compare how learning and memory change in various conditions. Scientists can compare how learning and memory are impacted during various disorders or during brain development using these mazes. Research in learning and memory can help determine how to minimize the changes in the brain that are observed in disorders in hopes of alleviating negative symptoms of the disease state.

One step inward from the macroscale of behavior is studying the whole brain and how it is connected (connectomes). Scientists who are working on this level are examining the connectivity of the brain. In learning and memory, scientists are trying to determine which circuits are necessary for learning and memory. For this, scientists are using computer modeling as well as animal models. A lot of work in this area have been completed in an animal model, *Caenorhabditis elegans*, which is the first animal model to have its entire genome mapped (20,000 genes) as well as its entire connectome (thinking cell connections) mapped. *C. elegans* is a nematode or roundworm. In total, it has a little more than 1000 cells, approximately 300 of which comprise the nervous system of the animal. The human connectome will take a little longer. Humans have a nervous system with billions of cells and 80,000 genes. The human connectome project is well underway. There is a lot of effort being made to complete the human connectome because it will allow us to better understand how changes in one region (good or bad) impact other regions (good or bad) (Fig. 5.3).

There are in fact several different "omes" that scientists study in relation to the brain. As we just talked about, there are connectomes, which examine the connections of the nervous system of a given organism. There are also genome studies, which examine the genome, or total genes of an organism—genome-wide analysis studies (GWAS it is pronounced with the letter "G" then the word "WAS" with a short A sound). With learning and memory, there are GWAS studies exploring which genes are turned on or turned off during learning and memory and how those are impacted during various disorders. There are also studies on total proteins (proteomes), total mRNA (transcriptomes), or metabolites (metabolomes) that are present in a particular organism at a given time in development. These last three "omes" are time dependent. Different genes are turned on or off during development and that impacts proteomes, trancriptomes, and metabolomes. With learning and memory, scientists are studying these different "omes" to determine which genes are expressed during learning and memory, as well as which mRNA are altered, which proteins are altered, and which metabolites are altered.

Figure 5.3 Computer reconstruction showing some of the connections in the human brain.

They can utilize a model system, ask that model organism to learn a task and compare the "omes" of the organism that learned a task to an organism that did not. This allows scientists to understand which part of all the "omes" are necessary for learning and memory.

The microscale end of the learning and memory research includes investigating a specific region, a specific type of cell, or a subcellular function. Regions of the brain could vary in size, or what types of cells that are present, what genes are regulating the region, how behavior generated from that region is altered or how the cells get to that region during development. Scientists who study the brain cells or inside the brain cells are examining the number of the cells present in a region, the way the cells communicate, the way the cell makes proteins or throws away trash. There are many types of cells and proteins that are necessary for learning and memory. As discussed earlier, learning and memory occur with channels moving into the receiving structure (dendrite).

Scientists combine the macroscale and the microscale with different model systems to get the best picture of what is going on in the brain at varying levels to better understand it's function.

5.3.1 Hippocampus History Lesson

What type of research was necessary to understand that the hippocampus was the region necessary for forming new memories? This discovery came through a series of unfortunate events. Among the neuroscience community, there is a very popular patient with a normal IQ, H.M., who was studied intensely by Brenda Milner. H.M. may be the most studied patient in neuroscience. Brenda Milner was finishing her doctoral research in 1952 at McGill University under the direction of her mentor, Dr. Donald Hebb, when she was invited to study patient H.M.

Patient H.M. was knocked down by a bicycle early in his childhood, resulting in minor seizures initially, and major seizures during the teenage years. In his late 20's, he had major seizures and was incapacitated by them to the point where he could not work or have normal life. The neurosurgeon in Hartford, William Scoville, recommended to remove part of his brain due to uncontrollable seizures. The part of the brain the doctors removed included the hippocampus. After the surgery, patient H.M. could not recall any new events or new facts. He could remember things prior to the surgery. When William Scoville had learned about the observations Brenda Milner had made in similar situations, he invited her to come study patient H.M.

The science gained from behavioral studies on H.M. analyzing his memory allowed neuroscientists to conclude that the hippocampus was a brain region necessary for the formation of new memories but not necessary in the recall of old memories. Patient H.M. could learn perceptual-motor (procedural) tasks, such as mirror tracing even though episodic memory was impaired. These studies indicated that the hippocampus was not involved in procedural or motor task learning.

5.4 HOW AGING IMPACTS LEARNING AND MEMORY

While many regions of our brain may still be maturing, our brain is actually at its peak weight around 18–20 years old. After that, the brain begins to gradually shrink. Those all important connections (synapses) that thinking cells (neurons) make in order to talk to each other, start to break down. And when synapses fail, neurons retract the parts of the connections (synapse) so there is no longer a connection(synapse). This death of connections (synapses) is what ultimately creates a loss of memory as we age. As we age, we are loosing the connections (synapses) that are necessary in order to form new memories and because of that, we become more and more forgetful.

Normal shrinkage in the brain is due to loss of connections (synapses). In persons with dementia, there is not only a loss of connections (synapses), but also death of some of the thinking cells (neurons) within the brain. Compared to a healthy individual of a similar age and gender, a brain from a person with dementia is smaller due to the loss of both connections (synapses) and thinking cells (neurons). And because there is a loss of both connections (synapses) and thinking cells (neurons), the impact on learning and memory is greater that loss of connections (synapses) due to normal aging process.

5.5 WHAT DISORDERS IMPACT LEARNING AND MEMORY?

There are several disorders that impact learning and memory. Much research is exploring how we can improve our learning and memory. These disorders include ADD, dementia, and Alzheimer's disease.

5.5.1 ADD/ADHD

Attention deficit disorder (ADD—each letter said individually) or attention deficit hyperactivity disorder (ADHD—each letter said individually) have no known genetic cause. There are a lot of debates surrounding ADD/ADHD. Some argue that it is over-diagnosed in the United States. Others argue that the increase in the number of cases is due to better diagnostics and increased awareness. Others argue that it is a change in our culture thanks to microwaves, fast food, and cable internet, all of which shrink our attention span.

Changes observed in patients with ADD or ADHD in the brain result in the loss of attention control, hyperactivity, or impulsiveness when compared to the nondisorder state. These symptoms are detected around 6—12 years of age. Most cases do not persist into adulthood, but for those cases who do persist, it does not impact lifespan. Studies have explored several changes in the brain between the disorder and the nondisorder state, looking at several chemical messenger systems pathways that are different but to date there is no known cause or cure. Treatment can include medication and/or behavioral therapy. ADD and ADHD are disorders that impact memory and that require more research and insight to better understand mechanisms altered in the disease.

5.5.2 Dementia

Dementia is a broad term that refers to a decline in total cognitive ability, particularly learning and memory, to a point that impacts daily function for a person. This decline is greater than what would be anticipated due to aging. Other diseases that affect the brain have increased risk for dementia or dementia may even be a common symptom of some disorders. Dementia can result in emotional problems, decline in language understanding, or a decrease in motivation. Dementia is more common in people 70−75 years of age. Complications that arise due to dementia may shorten a person's lifespan. There is no known cause and treatments usually target any disease symptoms.

5.5.3 Alzheimer's Disease

Alzheimer's disease (AD−letters said individually) is the most common form of dementia. There is no known genetic cause for AD. In the brain, there is a gradual loss in problem solving, memory, language, behavior, emotional function, and judgement. Some people develop hallucinations and/or delusions. Mild depression and social withdrawal are common among people with AD. During the initial stage of AD, which can last 2−4 years, a person with AD will have frequent memory loss which affects recent memories while older memories are preserved. Usual language is less fluid and language comprehension may begin to decline. Common objects may be moved to odd places and mood swings may occur frequently. In the second stage, which lasts 2−10 years, people with AD cannot mask their memory problems. Both recent and older memories fade away including names of family members or friends. Speech may not make sense. It is also common that people in the second stage of AD are easily lost and may not be able to find their way home. Social norms are forgotten and there is a lack of inhibition. There may be difficulties dressing oneself, eating, or even walking. In the final stage, which lasts 1−3 years, people with AD are confused and have difficulty carrying on a conversation. They have problems swallowing and mobility difficulties.

There are several documentaries and movies that demonstrate what AD is. HBO has a series of films called *The Alzheimer's Project* looking at what it is like to live with AD as well as the current scientific discoveries in AD.

There isn't a clinical test to confirm AD. There are tests that can rule out other disorders, but no clinical diagnosis can confirm AD. A diagnosis of AD can only be confirmed by autopsy. The brain of people with AD have a buildup of trash (called beta-amyloid plaques and neurofibrillary tangles) inside the cell. The plaques and tangles are not suppose to be there and results in unhealthy thinking cells (neurons) near the plaques and tangles, which results in an untimely death of those thinking cells (neurons). More research is necessary to determine if the tangles and plaques cause the disease or are a result of the disease.

Because AD is a progressive disorder, there is no cure. Treatment therapies target symptoms of the disease. Medications try to improve memory and also target depression which usually develops in patients with AD. There are also medications that can target hallucinations or delusions if those are symptoms reported.

5.6 CONCLUSION

We know that there are two main types of memory: explicit and implicit and how various channels must be increased in the receiving structure (dendrite) in order to form memories. We explored learning and memory research, as well as historical experiments that shed light on the role of the hippocampus during learning and memory. We discussed disorders that impact learning and memory.

We discussed the fact that normal aging reduces the number of connections (synapses) between thinking cells (neurons). Can we do anything to keep our brain sharp as we age? There are many games out there that boast that they can improve your learning and memory. Most of the scientific studies that have examined these games do not look at long-term effects, and the few studies that have looked at effects several months out from learning the task have mixed results. Games aside, there are everyday activities we can incorporate into our schedules that can improve our ability to think clearly, learn, and remember. We have already discussed that sleeping and eating will help improve learning and memory as well as daily exercise. With those, activities that require attention and focus as well as taking in new information or problem solving are great for helping out your overall ability to learn and remember.

BIBLIOGRAPHY

Baars, B., Gage, N., 2012. *Fundamentals of Cognitive Neuroscience*. Elsevier Academic Press.
Kandal, E.R., Schwartz, J.H., Jessel, T.M., 2000. Principles of Neural Science, fourth ed. McGraw-Hill Companies.

Kolb, B., Whishaw, I.Q., 2014. An Introduction to Brain and Behavior, fourth ed. Worth Publishing.

Mason, P., 2011. Medical Neurobiology. Oxford University Press.

Nicholls, J.G., Martin, A.R., Fuchs, P.A., Brown, D.A., Diamond, M.E., Weisblat, D., 2012. From Neuron to Brain, fifth ed. Sinauer Associates, Incorporated.

Purves, D., 2012. Neuroscience, fifth ed. Sinauer Associates, Incorporated.

Sontheimer, H., 2015. Diseases of the Nervous System. Elsevier Academic Press.

Squire, L., Kandel, E., 2009. Memory: From Mind to Molecules, second ed. Roberts and Company Publishers.

Squire, L., Berg, D., Bloom, F., du Lac, S., Ghosh, A., Spitzer, N., 2012. Fundamental Neuroscience, fourth ed. Elsevier Academic Press.

CHAPTER 6

11:30 a.m. Hanger: (n) Hunger-Induced Anger

Contents

SUMMARY

In this chapter, we are going to talk about the motivation underlying certain behaviors, such as eating, and how the brain utilizes food to gain nutrition necessary for function. What happens when that system in the brain becomes addicted.

6.1 HUNGER

How does our body know we are thirsty? There are sensors in the brain (hypothalamus) that monitor the levels of water in the body. These sensors give feedback to the hypothalamus, which can then use that information to generate a signal that lets us know if we are in need of water. In order for our body, which is 70% water, to function properly, the hypothalamus must monitor the water levels inside our body. By way of reminder, the hypothalamus is part of the diencephalon that regulates stress, eating, growth, drinking, and maternal behaviors.

And how do we know we are hungry? There are sensors in the hypothalamus that monitor the nutrients within the blood. When we get low

Neuroscience Basics.
DOI: http://dx.doi.org/10.1016/B978-0-12-811016-4.00006-4

on nutrients, the hypothalamus interprets that signal and will create a hunger signal. When nutrients are low, "hanger" happens. Hanger is hunger-induced anger or irritability. Based on what nutrients are lacking, the hypothalamus could generate two types of signals—an overall hunger signal or a more specific craving. The cravings we experience can be due in part to nutrients that the body needs. Cravings can also be due in part to changes in hormones and stress because the hypothalamus also regulates the stress levels and thus the craving is usually for a food that bring pleasure or comfort to eliminate the stress. The brain signals the need for water or nutrients, we consume them, and then, the body reports back to the brain that all is well.

There are other brain signals that occur when we eat. There are different types of taste buds on our tongues responsible for sensing salty, sweet, and savory. And the combination of flavors, fats, carbohydrates, and sugars in our food communicates different messages to our brain. The signals regarding food or hunger can be coupled with memories or experiences while eating. But the drive to avoid hunger and dehydration are hard-wired into the hypothalamus for our survival.

6.2 BLOOD–BRAIN BARRIER

How does the brain send those signals? Or how does it know when the food is finally been consumed? The brain knows when the nutrients are present because of the nutrients in the blood. The nutrients get into the brain by crossing the blood–brain barrier (BBB—each letter said individually). The BBB works to keep out harmful materials while allowing in the necessary molecules found in the bloodstream. If harmful chemicals attack the skin cells, while not pleasant, the end result is usually not horrific because skin grows back. The problem with harmful materials getting into the brain is that thinking cells (neurons) do not usually regenerate. Death of thinking cells (neurons) in certain areas of the brain could result in loss of mobility or loss of memory, etc. So the BBB protects the brain from harmful substances.

One of the types of supporting cells (glia), astrocytes, has a special task with the BBB. Astrocytes form a special structure on blood vessels called end feet. These structures exist so that astrocytes can dump any unwanted chemicals or molecules from the brain into the blood for recycling or waste removal. Some of these unwanted chemicals or molecules are excess of signaling molecules. Clearing the extra from the brain makes sure that the brain keeps a healthy balance of the signal. The end feet also allow the astrocytes to take in glucose—the food of the brain.

While the BBB does an excellent job of keeping harmful materials out of the brain, it also makes it difficult to design drugs that target the brain. Some drugs for neurological disorders have great promise but cannot cross the BBB. For example, the main pharmacological treatment for Parkinson's disease, L-DOPA (say the letter "L"—and then Do-pa is said as a word), is the precursor to the neurotransmitter dopamine. What the brain needs is dopamine. But dopamine does not cross the BBB and L-DOPA does. So the current treatment is the precursor not actually the chemical needed because of the BBB.

6.3 BRAIN FOOD

The brain receives the majority of its oxygenated blood from the two common carotid arteries on either side of the neck, which branch into the internal carotid artery and external carotid artery. The external carotid artery is responsible for transporting the majority of the blood that supplies the neck and the face. The internal carotid artery is responsible for transporting about 75% of the total blood flowing to the brain. The internal carotid artery divides into the middle cerebral artery and anterior cerebral artery. The middle cerebral artery is the larger of the 2 branches and it further branches into 12 smaller branches. Those 12 smaller branches are responsible for providing blood to the frontal, temporal, parietal, and occipital lobe. To meet the needs to the brain, approximately 0.5 L of blood are pumped through the brain each minute.

In order for the brain to have more cells, thinking cells (neurons) gave up the ability to store glycogen. Glycogen is the main long-term way that our body stores glucose. The other way the body stores energy is in fat cells. Glycogen stores in the body are dependent on physical activity, basal metabolic rate, and eating habits. Glycogen is made and stored mostly in the liver and muscle cells, although there is some glycogen stores in the kidneys, trace amounts in supporting cells (glia), white blood cells, and in the uterus during pregnancy. When glycogen in the liver is broken down to glucose, that glucose is used throughout the body, including the brain. So diet and exercise impact glycogen stores and that impacts the amount of glucose in the brain. And the brain needs glucose to function because it doesn't have its own warehouse of glycogen.

How much glucose does the brain use? In an adult, the brain accounts for approximately 2% of the total body mass, but the brain utilizes approximately 20% of the total energy requirements. Energy requirements for a child's brain which is still developing is approximately 40% of total energy requirement. The adult brain requires about 150 g of glucose a day. The adult brain also requires about 72 L of oxygen each day. The brain therefore must have a supplier in order to get oxygen and glucose across the border—the BBB. Glucose and oxygen are trafficked to the brain and can get across the barrier through the bloodstream.

6.4 WHY DOES THE BRAIN WANT US TO EAT?

Okay, so we know why we get hungry or why we get thirsty, but why do we want to eat? How does our body know that eating will solve the hunger situation? There are really several different behaviors that fall under this section, not just eating. We are about to walk through how the brain processes pleasure. While not all food is pleasurable, the feeling of being full is pleasurable. This is a good thing that keeps humans from going hungry.

The brain developed a few regions that give feedback for behaviors in order to keep us alive and keep the human race going. We have a system in our brain that gives us gold stars for doing things that will keep us alive. Imagine with me thousands of years ago before gastro pubs and farm to table initiatives were the thing. This is before we knew anything about fats or gluten or how many calories a day is considered normal, before restaurants and grocery stores. When our early ancestors consumed food, the full feeling sensed by stomach was communicated to the brain, and the brain released a chemical message (the neurotransmitter—dopamine) in the reward pathway (nucleus accumbuns (Nacc) and ventral tegmental area (VTA)) that communicate satisfaction (Fig. 6.1). The same region (the reward pathway) releases the same chemical message (dopamine) after sex. Food and sex are evolutionarily beneficial. They keep the individual thriving and the species thriving. Behaviors that are beneficial to us or to the species naturally trigger a release of satisfaction in the reward pathway (Nacc/VTA). Doing great on a test, getting a job, getting a promotion, getting a car, winning a game, etc. result in positive feedback in the reward pathway (Nacc/VTA). Other behaviors can trigger displeasure or repulsion in the reward pathway, which again can be evolutionarily beneficial because they keep us from repeating the behavior (usually).

Like so many of the chemicals messages (neurotransmitters) in our brain, there is a normal level (homeostasis). Our brain keeps a tight regulation on how much of that chemical signal is actually used to communicate. Too much or too little of the chemical message (neurotransmitter)

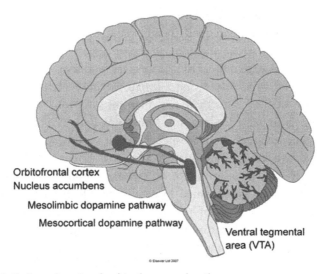

Orbitofrontal cortex
Nucleus accumbens
Mesolimbic dopamine pathway
Mesocortical dopamine pathway
Ventral tegmental
area (VTA)

© Elsevier Ltd 2007

Figure 6.1 Brain regions involved in the reward pathway.

in the reward pathway (Nacc/VTA) is not a healthy situation and can completely alter a person's motivational states.

6.5 WHAT HAPPENS WHEN THERE IS TOO LITTLE BLOOD OR TOO MUCH DOPAMINE?

There are several brain disorders that are a result of the damage to the brain regions themselves through force or impact or damage to the blood flow. We will discuss how stroke impacts brain function. We will also discuss how the reward system can be altered, resulting in addiction.

6.5.1 What Happens When Glucose and Oxygen Cannot Get to Cells?

What happens if the blood flow through the brain is disrupted, and glucose and oxygen cannot get to the brain cells? There are three types of stroke (cerebrovascular infarct or CI is the technical term—each letter said individually) which can block the blood flow: blood vessel blockage (focal ischemic stroke), global blood flow reduced (global ischemic stroke), or a rupture of particular blood vessels (hemorrhagic stroke), which results in reduced blood flow to the brain. This results in reduced amount of glucose and oxygen to an affected area of the brain. It also reduces the ability of the brain to clear the excess chemicals, which can be toxic to the cell and potentially neighboring cells.

Risk factors for stroke include age, heart disease, diabetes, smoking, and drug use. Because the risk factors are additive, controlling risk factors is imperative to preventing a stroke. Because a stroke can happen in any region of the brain, there are different symptoms based on what brain region is impacted. Common symptoms include changes in consciousness, headache, changes in speech, facial weakness, uncoordinated movement or weakness, and poor balance.

Current treatment of stroke is to restore blood flow to the area. Usually, a brain scan is performed to determine which type of stroke has occurred. If stroke is a result of blocked blood vessels, the clot is broken up through the use of a catheter or a pharmacological agent (tPA—each letter said individually). If the stroke is a result of a rupture, repair to the vessel is necessary to stop the bleeding in the brain and restore normal blood flow.

My Beautiful Broken Brain (2014) is a documentary of a hemorrhagic stroke victim which depicts life of a patient following a stroke.

6.5.2 When the Reward System Is Hijacked: Addiction

When we release too much of the satisfied message (dopamine) in the reward pathway (Nacc/VTA), our brain starts to have problems. Say my brain is use to 1 L (completely arbitrary) of the satisfaction message (dopamine) in the reward pathway (Nacc/VTA) and that is normal for me. Then a behavior or a substance ups that level so now my brain thinks that 100 L of the satisfaction message (dopamine) is normal. If the behavior or the substance is repeated, then the reward pathway (Nacc/VTA) gets use to having 100 L of the satisfaction message (dopamine) instead of 1 L.

So when the substance or the behavior is no longer present, the 100 L of the satisfaction message (dopamine) is no longer there and the 1 L level is present. When that happens, the brain feels the difference and determines it has to do something to get back to the 100 L of the satisfaction message (dopamine). It feels like 100 L is now normal and 1 L is no longer normal and it is way too low. So the brain is driven to get to 100 L *despite any adverse consequences.*

This is how the reward system gets hijacked during addiction. In studies with primates, the results demonstrate that the release of the satisfaction message (dopamine) occurs with a cue that the behavior or substance is about to come, not during the actual behavior or consumption/administration of the substance. The reward is triggered in the brain before the substance or behavioral side effects ever hit the brain. The cue for the addictive substance is just as powerful as the actual substance itself. This is one of the many reasons why addictions are difficult to overcome. The brain is literally rewired and trying to reestablish "normal" in the absence of the substance or behavior is difficult, especially if there are cues for the substance.

How many times does someone have to use a substance or repeat a behavior in order for it to rewire the brain and form an addiction? That is different for different people because of genetics. Interestingly, evidence is starting to suggest that there may be some genetic predisposition to addiction. Think about what we have mentioned so far in regards to the reward pathway. We know that there are channels that receive the chemical message (dopamine) and that there is chemical released to communicate pleasure or displeasure. And our brain has a perception of the amount of "normal" chemical present during a pleasure response. Genetics can alter any of these key players in the reward pathway making a person more susceptible to addiction. That doesn't mean that genetics make an addict. It does mean that genetics may make a person more vulnerable to addiction. And how the brain is wired to begin with determines how much exposure is required to form an addiction.

There is a difference between addiction and dependence on a substance or behavior. We may be dependent on a routine or caffeine, but wouldn't necessarily pursue those despite adverse consequences. Addiction is pursuing whatever upped the satisfaction message (dopamine) despite harm and potentially at great personal costs.

There are a great many substances and behaviors that can result in addiction. Some of the substances mimic the natural chemical messages (neurotransmitters) in the brain. The reaction of those knock-off chemical messages (neurotransmitters) results in some action in the brain which can be rewarding. Remember when we were talking about exercise and how the endorphins released are the natural chemical and morphine is the knock-off made to act like endorphins? Well, morphine talks to the receptors that endorphins are supposed to be talking to. And morphine produces a sedative effect the way endorphins do. So the overall experience can be a rewarding one, which can create an addiction to morphine.

The most successful treatment of addictions is usually a combination of medication and behavioral therapy. Addiction impacts teens to older adults. The number of persons with addictions is probably grossly underestimated due to the fact that most persons with addictions do not seek treatment because they do not believe they have a problem or do not want to be plagued with the social stigma of addiction (Fig. 6.1).

There are many, many documentaries and movies about the life of a person dealing with addiction. *Dope Sick Love* (2005) and *Overtaken* (2010) are just a few of the hundreds.

6.6 CONCLUSION

Detection of water in the body and nutrients in the blood allow the hypothalamus to sense if we are hungry or thirsty. Blood protects the brain, eliminates waste, and delivers the much needed nutrients of glucose and oxygen. Feedback from healthy habits signals pleasure in the reward pathway to keep us alive and the species thriving. As humans, maintaining healthy blood flow to the brain is necessary for proper thirst and hunger regulations as well as to avoid stroke. A healthy lifestyle includes not smoking, a healthy weight, and not using drugs.

Finally, this chapter covered addiction. Enter in our topic of love. An interesting conversation among neuroscientists is whether or not love is an addiction. As it currently stands, in the large manual that describes mental disorders (diagnostic and statistical manual of mental disorders (DSM), version 5), romantic love is not listed.

Here is some of the current debate about love as an addition. The definition of addiction is some behavior or substance that increases the satisfaction message (dopamine) to a level that results in a person who will pursue behaviors or substances that keep that satisfaction chemical high despite potential harm. Often humans pursue love with the full knowledge that we may have our hearts broken either through a fault in character or through an accident. We accept these conditions and still seek out a partner with whom to share our life journey. Vows exchanged in marriage echo this commitment "in sickness and in health, in want and in plenty." In love, we pursue a person despite adverse consequences. Further, some scientists have reported that the brains of broken-hearted individuals are similar to those of recovering addicts. The broken-hearted have cravings for the person who use to be there, potentially to negative ends. So, is romantic love an addiction? It is an interesting debate that continues.

BIBLIOGRAPHY

Augustine, J.R., 2007. Human Neuroanatomy. Elsevier Academic Press.
Baars, B., Gage, N., 2010. Cognition, Brain and Consciousness: An Introduction to Neuroscience, second ed. Elsevier Academic Press.

Diagnostic and Statistical Manual of Mental Disorders. 2013. fifth ed. American Psychiatric Association.

Kandal, E.R., Schwartz, J.H., Jessel, T.M., 2000. Principles of Neural Science, fourth ed. McGraw-Hill Companies.

Kolb, B., Whishaw, I.Q., 2014. An Introduction to Brain and Behavior, fourth ed. Worth Publishing.

Mason, P., 2011. Medical Neurobiology. Oxford University Press.

Nicholls, J.G., Martin, A.R., Fuchs, P.A., Brown, D.A., Diamond, M.E., Weisblat, D., 2012. From Neuron to Brain, fifth ed. Sinauer Associates, Incorporated.

Purves, D., 2012. Neuroscience, fifth ed. Sinauer Associates, Incorporated.

Sontheimer, H., 2015. Diseases of the Nervous System. Elsevier Academic Press.

CHAPTER 7

1:00 p.m. Mid-Afternoon Blues

Contents

SUMMARY

The brain is able to generate moods, emotion, and behavior. Are we the only animals that have moods and behavior? We can answer this in part based on our classification system (taxonomy) and in part based on our evolution. How did behavior drive our evolutionary process? What are some moods, behaviors, and human emotions? What are disorders that result from a disruption of emotions?

7.1 HUMAN MOODS AND HUMAN BEHAVIOR

In the early 1800's, the study of phrenology was a hard science. Phrenology comes from the Greek: *phren* meaning "mind" and *logos* meaning "knowledge." Phrenology describes an enlargement of local areas of the brain that produces characteristic bumps and ridges on the skull and studying these bumps and ridges could determine a person's character. Phrenology was developed in the late 1700's by a German physician, Franz Joseph Gall. It was popular from early-1800's until

Neuroscience Basics.
DOI: http://dx.doi.org/10.1016/B978-0-12-811016-4.00007-6

mid-1800's. While phrenology is no longer regarded as science, Gall's hypothesis that character, thoughts, and emotions were localized to specific regions of the brain is an important advance in neuroscience.

To better understand human behavior and emotions, we have to understand how we are different than other organisms and human evolution.

7.2 TAXONOMY—THE CLASSIFICATION OF ORGANISMS

Humans have an amazing brain that couple both thought and movement to generate behaviors. But do all living organisms have such a complex organ? The answer to the question is found in taxonomy, which is a science used to define biological groups. Carl Linnaeus (1707—78) was a Swedish botanist and is regarded as the father of taxonomy. Among his many contributions to science and taxonomy, he formalized the binomial nomenclature—the system by which a living creature is given two names—the genus and the species in Latin.

Taxonomy is from the ancient Greek: *Taxis* means "arrangement" and *nomia* means "method." Therefore, taxonomy is the science and study of classifying creatures based on shared characteristics. These shared characteristics could be genetics or morphology or an organisms role in an ecological setting. Depending on which set of characteristics are used, organisms may be classified in different ways. To that end, frequently there are debates on the classification of some living organisms.

Each living thing is classified by kingdom, phylum, class, order, family, genus, and species. Each of those classifications give more and more specificity moving away from kingdom and toward species (Fig 7.1). Classifications give

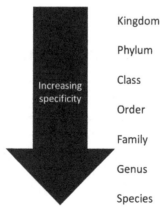

Kingdom

Phylum

Class

Increasing specificity

Order

Family

Genus

Species

Figure 7.1 The levels of taxonomy.

information about an organism—is it made up of one cell or many cells? Does it give birth to live young or does it lay an egg? Scientists can tell quite a bit about a living thing based on what classification it is given.

Understanding the brain and its evolution in part means understanding why the human brain is different from other creatures on this planet. To better understand these differences and the uniqueness of the human brain, the following section describes the classification of humans.

The five kingdoms are Monera (mostly bacteria), Protista (single cell organisms like algae), plantae (plants), Fungi (fungus), and finally animalia (animals, including humans) (Fig. 7.2). Of all the five kingdoms, only Animalia evolved with muscles *and* a brain of some sort. With a brain and muscles, there are a great many things we can do that other living organisms cannot accomplish.

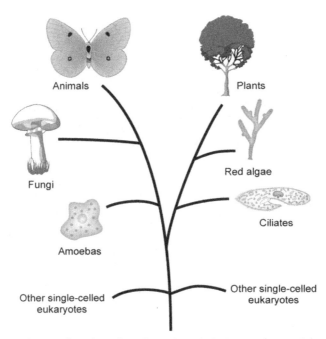

Figure 7.2 *Eukaryote kingdoms based on ribosomal RNA sequences.* Comparing the ribosomal RNA sequences has allowed scientists to deduce how closely the major divisions of eukaryotic organisms are related. A variety of single-celled eukaryotes were the earliest groups to branch from the ancestral eukaryotic lineage. As can be seen, the fungi are more closely related to animals than to plants. The branch leading to higher plants includes several groups of algae, including the red algae, as shown here.

7.2.1 Human Classification

7.2.1.1 Kingdom Animalia

The animal kingdom has over a million species with a large variety of complexity. Starfish, insects, sheep and lightening bugs fall into the animal kingdom with humans! One of the shared characteristics among the animals is that nerve cells and muscles are used together for locomotion and behavior.

7.2.1.2 Phyla Chordata

Animals in the same phyla as humans include frogs, sharks, birds, and dogs. Phyla chordata have a spinal cord. Other shared traits include bilateral symmetry meaning you can cut a chordate in half and both sides are a reflection of the other and the fact that the spinal cord and brain are enclosed in bone. This makes sense because all that information needs protection. Animals in older phyla have what is called a nerve net, which has cells that receive sensory information and those cells connect directly to cells that move the muscle. Nerve nets are typical in jellyfish and sea anemones.

7.2.1.3 Class Mammalia

Mammals are warm-bodied animals that have hair on their body and produce milk to feed their young. One of the shared characteristics among mammals is a large brain size and social behavior. Other animals in this class are kangaroos, cows, primates, and dolphins. So even at this level of specificity, there is still quite a bit of diversity.

7.2.1.4 Order Primates

The order of Primates includes a little over 200 species. Primates have excellent vision with forward-facing eyes which are close together, and a reduced sense of smell compared to other mammals because of the increased use of vision. Primates are also dexterous and have a collar bone to allow greater mobility in the shoulders compared to other mammals. And most importantly, primates have larger brains than a nonprimate mammal with a similar body size. In this case, bigger is better.

7.2.1.5 Family Homindae

Homindae is also known as the great apes or hominids. This family includes seven species: the Borean and Sumatran orangutan, Gorilla (both east and west), the common chimpanzee, bonobo, and humans. And here is an example of one of the debates on animal classifications. Scientists debate what species are part of the Homindae family. Currently, hominid refers to the great apes including the humans. In previous years, the hominid classification was specific to only humans. And it may change yet again in the future.

The end of the human taxonomy—Genus *Homo* and species *sapiens* (which includes the genus in the actual name of our species)—*Homo sapiens*. In Latin, *Homo* means "human" and *sapiens* means "wisdom." Species name, *Homo sapiens*, means "wise human."

Human taxonomy builds the foundation to understand how human evolution directed current human brain size. As phyla chordata (animals with spinal cords) were evolving, complex behaviors like improved learning, highly organized social behavior, and eating with our hands were also evolving.

Understanding how humans are classified and the fact that we are talking about tissue in our skull lays is the necessary background information to fully appreciate the human brain evolution.

7.3 HUMAN EVOLUTION

Current human brains are somewhat of a recent development in evolutionary history. The first human-like brain evolved around 6 million years ago and was a little less than half the size of our current brain. The more modern human brain (something similar to what is interpreting the symbols on this page for you) evolved around 200,000 years ago (Fig. 7.3).

Chronological order of human evolution:

Homo habilis: 2.8 million years ago up until 1.5 million years ago

Homo erectus: 1.9 million years ago up until 70,000 years ago

Homo neanderthalensis: 300,000 years ago up until 28,000 years ago

H. sapiens: 200,000 years ago to present

H. habilis remains, the oldest fossils designated in the human genus (*Homo*), were discovered in Olduvai Gorge in Tanzania by a team led by Louis and Mary Leakey during 1960—63. The species was designated *H. habilis*, which in Latin means "handy human", to signify that these humans were tool makers. *H. habilis* fossils have been extracted in Tanzania and Kenya. *H. habilis* had a brain volume of about 640 cm^3. The designation in the genus *Homo* have been a subject of controversial debates ever since the discovery of the skeletal remains. Some taxonomist argue that *H. habilis* does not belong to the genus *Homo*, but to genus *Australopithecus* as *Australopithecus habilis*.

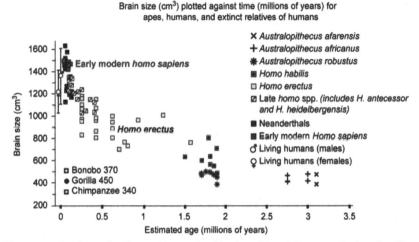

Brain size (cm³) plotted against time (millions of years) for apes, humans, and extinct relatives of humans

Figure 7.3 Evidence for the rapid growth of brains of hominine over the last 2 million years. The brain sizes of modern chimpanzees and gorillas have been added for comparison. *Modified from McHenyr, H.M., 1994. Tempo and mode in human evolution. Proc. Natl. Acad. Sci. 91 6780–6786.*

H. erectus remains were discovered in 1891 by a Dutch anatomist, Eug'ene Dubois. *H. erectus* in Latin meaning "upright human" (named on a wrong notion that they were the first upright human) demonstrate a considerably larger brain volume of approximately 1200 cm³, almost 2 times that of *H. habilis*. Amazing what 1/2 million years can do, isn't it?

H. neanderthalensis were named after a region in Germany where the first skull was found, present day Belgium. Philippe-Charles Schmerling (1791–1836) discovered the first Neanderthal fossil, a partial cranium of a child. *H. neanderthalensis* fossils have been found in Western Europe, Central Europe, the Balkans, Ukraine, Russia, and parts of Asia. *H. neanderthalensis* had a larger brain than *H. erectus* with a volume of approximately 1600 cm³. It is known that *H. neanderthalensis* had a hunting culture, wore jewelry and makeup and overlapped in existence with *H. erectus*.

Homo sapiens (meaning "wise human" in Latin) appears within the fossil record about the last 200,000 years. The modern brain is approximately 1400 cm³. Genetic evidences supports the hypothesis that *H. sapiens* have a recent, and single origin. According to both the genetic evidence and the fossil record evidence, *H. sapiens* evolved in Africa between 200,000 to 100,000 years ago.

The taxonomy and evolution of the current human brain give the background necessary to understand brain function. But how did the current brain evolve?

7.4 WHAT WERE THE DRIVING FORCES FOR HUMAN BRAIN EVOLUTION?

Muscles and the brain evolved and developed in a similar time frame for organisms classified as animals, which makes sense because working together, the muscles and the brain generate complex behaviors. Current human brains developed at the same time as the muscles so that the muscles had some form of control or governing machinery. The brain's primary function is to generate a behavior. In order to do that, the brain has to gather information about the world, integrate that information about the environment, and produce commands for the muscles to carry out.

Why did our brains and muscles evolve together? Why is the primate brain so complex and interesting? There are a few theories scientists are currently exploring to explain the why and the how of the coevolution of the brain and muscles to generate behavior.

1. Primates need to forage for food which requires complex behavior.
2. Increased skull space increased the blood flow.
3. Slow personal development allows for more brain cells to be made.

The first theory states that our current human brain is so complex because of what is required in order to forage. Primates (humans are in the Order Primate) are foragers. Animals that graze simply need to walk around to find their next meal. Animals that forage need to remember where food sources exist. At least, those that want to live will need to remember where the best food sources are located. Those primates will also need fine motor skills to access the food. If it is small berries, they will need to be able to have appropriate depth perception to understand how far away the berry is, proprioception (understanding of where their body is in space), the ability to extend an appendage an appropriate distance, the ability to pinch/grasp the berry, and finally move the berry to the mouth. Who knew berry picking was so complex? The first theory concludes that modern brains are more complex because more complex behaviors are required in order to successfully forage.

The second theory states that modern brains are more complex because modern humans used their brains more, thus generating heat, resulting in increased blood flow, which allows for more growth. The brain needs blood flow for growth. The brain uses blood to clear out extra stuff which could become toxic and blood also brings in nutrients. This second theory states that the more the brain was used for tasks, the blood flow increased, which allowed for growth and development of a bigger brain.

The third theory states that humans are all slow. Humans mature and develop slowly. In comparison to many other mammals (humans are in class Mammalia), humans are under parental care much longer many other mammals. Dolphins, which usually live to around 40 years, spend the first 3—6 years with their mothers, which is 7.5—15% of their life span. On average, humans spend almost the first 20 years with parental care, which is around 25% of their lifetime. This slow infusion of knowledge from the adults in the community means maturation and development are slower for humans. Because of this, the third theory states that modern human brains are bigger and can generate behavior because development is slow.

Each of these three theories are possible explanations to how our current brain evolved. It is probably the influence of all three theories that resulted in our brain evolution. Current studies are working to generate evidence which will determine which of these theories is most valid.

7.5 HUMAN EMOTIONS, BEHAVIORS, AND MOODS

With knowledge of our classification and evolution, we can better understand current human behavior. Our behavior is generated through connected thinking cells (neurons). Behavior can be something as simple as the knee—jerk reflex. When the tendon just below the knee cap (patella) is tapped, it will stretch the quadriceps femurs and cause it to contract. This produces the knee—jerk reaction. This is a simple type of behavior known as a spinal reflex. Spinal reflexes mediate behavior through the connection of thinking cells (neurons) all within the spinal cord. The movement is generated through those thinking cells (neurons) and does not involve the brain.

But other behavior generated by our brain takes into account our mood or our feelings or our emotions. As any of us who have gone through puberty know human emotion is not simple at all. Our feelings or emotions like anger, fear, sadness, jealousy, embarrassment, joy, happiness, or satisfaction are subjective interpretations that our brain makes. Because it is subjective, a certain scenario may bring joy to one person and sadness to another. Our life experiences shape how our brain interprets what we experience. Take fireworks. For some, they bring joy and awe and wonderment. For people who have experienced war time bombings, or small children who do not like loud noises, fireworks are terrifying. The brain is wired to aid survival. Small children associate loud noises with danger. People who have been in a war zone associate loud popping with gunfire. The brain's response to fire works for these groups exists as a way to protect life.

What behavior or emotions do we encounter in a day? We have two types of behaviors—regulatory and nonregulatory. Regulatory behaviors motivate individuals survival. We are motivated to eat when we are hungry. We grab a sweater when we feel cold. Nonregulatory behaviors are not necessary for survival. Examples are sexual behaviors, parenting, curiosity-driven investigations. As our frontal lobes evolved into larger structures, our range of nonregulatory behaviors increased as well.

Many of our emotions drive our behavior. For instances, when we are faced with moral dilemmas, often as humans, we make a decision based on what we "feel" is right and justify that decision with logic once we have made the decision. This is not always the case, however, often we do not consider rational arguments in rational decisions—we do what we "feel" is right.

So what influences our emotions and behaviors?

7.5.1 Nature versus Nurture

What drives our behavior? Is it our genetics or is it our environment? Is it nature or is it nurture?

The argument for nature is as follows. The brain is wired to keep an individual alive. The brain is wired from birth to have adaptive responses that will only aid in survival. As we have evolved, those individuals who have survived long enough to have offspring are well adapted and they pass those traits down to their offspring. In the whole nature versus nurture argument, these regions of the brain come prewired with an innate response for prosurvival. This means that initially, a lot of behavior is nature. These prewired prosurvival mechanisms in the brain can be modified through development. While the nature argument explains much, it does not explain many more complex behaviors.

The argument for nurture or the environment-driving behavior is as follows. Events or experiences in an individual's life function as a reward or a negative reinforcer. When a similar event or experience is repeated, the particular response (positive or negative) is likely to occur again. For example, the one time you got food poisoning was at a particular restaurant or eating a particular food. Even if the event was a fluke, we are most likely to not frequent that restaurant or order that particular food again.

Most neuroscientists argue that it is an interplay of both our genetics (nature) and the environment (nurture). Francis Galton (1822–1911) was one of the first scientists to explore the interplay of nature and nurture. In 1869, Galton published a book, *Hereditary Genius*, which explored the increase in incidence of gifted individuals who came from families with high mental abilities. He not only looked at the link between mental ability and heredity but also examined that most of these individuals shared social, educational, and financial advantages and that these environmental factors could also account for the correlation between these family members and the mental ability. In 1883, he developed a technique that remains in use today to determine if a factor is genetic or environmental—the twin study. Twin studies allow scientists to compare shared and nonshared components of the twin's lives and compare those variables to the fact that they have the same genetic make up. The complication for behavioral studies is that behavior is usually generated by several if not hundreds of genes not just a single gene.

Another consideration is why do we have emotions? What purpose do they serve? Certain emotions have a clear role. Fear and anger allow us to confront or run from danger. Distasteful experiences may prevent poisoning. Other emotions have a less clear evolutionary benefit for humans—happy, sad, embarrassment, etc. However, researchers are looking at potential explanations for emotions.

7.5.2 Brain Regions for Emotions

Behaviors are influenced in part by our emotions. Our emotions are influenced in part by our hormones that are released during our daily biorhythms (discussed a little more in the next chapter), in part by the environment, and in part by our past experiences. Our emotions are the chemical message in our brain, how we are sensing our environment, and our memories of the past. All of these are regulated by different regions of the brain that work together to regulate our emotions.

The brain area traditionally associated with emotions is the limbic system—which includes areas surrounding the thalamus such as the amygdala and the hippocampus (Fig. 7.4), both of which are involved in learning and memory. The amygdala is also involved with fear and emotions.

Emotional behavior requires different regions and structures for a typical emotional response. Say for instances that you are laughing heartily at an internet cat video. Usually when we laugh, a smile spreads across our

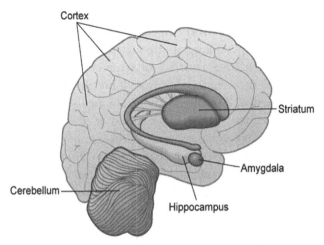

Figure 7.4 Regions of the brain involved in emotions.

face, so we are engaging muscles and nerves in our face. We make a noise of some sort when we laugh and so we engage our vocal chords. And the power for the noise comes from our diaphragm so we engage our lungs and diaphragm to create the power for the noise. And for this particular example, part of our brain recognized the cat doing something funny, so there was recognition of the element that made the situation humorous. All of that for a split second reaction of a laugh. While there are many key players necessary for different emotions and behaviors, these key players report back to or are driven by the hypothalamus, pituitary gland, the limbic system, and part of the cortex (memory storage) (Fig. 7.4).

7.6 DISORDERS WITH DISRUPTED EMOTIONS

There are several disorders with disruptions in emotions such as major depression, anxiety disorders, and controlled eating disorders. Mood disorders include depression and bipolar disorder, which we covered in an earlier chapter. So for this chapter, we will look at anxiety disorders and eating disorders. *Secret Fear* (1997) is a documentary that looks at the full spectrum of anxiety disorders—including phobias, panic attacks, and OCD.

Anxiety disorders are more common than depression. These disorders can include posttraumatic stress disorder (PTSD), phobias, panic disorders, social anxiety disorder and obsessive—compulsive disorder (OCD).

In our normal life, we experience anxiety. What defines these disorders is persistent fears or constant worry in the absence of direct threat. Often these disorders have physiological stress responses that are measurable. Anxiety disorders can have a variety of symptoms including feelings of panic or fear, problems sleeping, shortness of breath, chest pain, heart palpitations easily fatigued, difficulty in concentrating, or muscle tension.

In anxiety disorders, there are regions of the brain that appear overactive, which can result in several of the symptoms of anxiety disorders. One of the theories regarding anxiety disorders is that there are low levels of the main braking system for the brain—GABA. Remember that the chemical messages (neurotransmitters) tell the brain how to react. What if those messages had brakes that only worked part of the time but not all the time? The brain would be hyperactive. Low levels of GABA could create a situation where anxiety disorders would result. One of the important areas of the brain researched in anxiety disorders is the amygdala, which is central to processing fear and anxiety. Sensory information enters the amygdala and relates information about fear to other areas of the brain. In anxiety disorders, one of the main problems is persistent fear even though the threat has been removed. So communication breakdown potentially in the amygdala is an area of research scientists are investigating in anxiety disorders. Also being tested are areas that the amygdala communicates with to generate and control a fear response.

Anxiety disorders have no known genetic cause, but genetics may play a role in making a person susceptible to an anxiety disorder. There are several risk factors for anxiety disorders including: behavioral inhibition in childhood, having few economic resources, exposure to stressful events in childhood, anxiety in close biological relatives, and/or parental history of mental disorders. Anxiety disorders are usually treated with psychotherapy and medication.

People with panic disorders have a recurring and unexpected panic attack. Panic attacks are sudden intense fear that can increase the heart rate, increase perspiration, potential trembling, and/or shortness of breath. Usually, panic disorders are characterized by panic attacks, feeling out of control because of the unknown nature of the panic attacks, intense worry about when the next attack will occur, and potentially avoiding places where previous attacks have taken place.

People with social anxiety disorder, or social phobia, have a deep fear of settings where they expect they may feel embarrassed, judged, or rejected. These settings are often social settings or performance settings. Usually, social anxiety disorder is characterized by feeling highly anxious or self-conscious around other people, afraid people have already judged you, staying away from social settings, difficulty in making and keeping friends, or feeling nauseous when people are around.

PTSD is an anxiety disorder that results from experiencing a traumatic situation such as combat, natural disaster, rape, child abuse, bullying, a serious event occurring in the life of a loved one, or a serious accident. Common symptoms of PTSD include being hypervigilent, flashbacks to the traumatic situation, avoidance behaviors, anxiety, anger, bad dreams related to the situation, distress, and depression. These symptoms can last for months or years after the situation. Persons at increased risk for PTSD included people serving in the military, victims of natural disasters, victims of war-torn areas, and victims of crime. While uncertain of the exact brain alterations in people with PTSD, antidepressants that block the reuptake of serotonin are usually beneficial for people with PTSD, suggesting a potential role for serotonin. Therapy is also beneficial for people with PTSD.

OCD use to be classified as an anxiety disorder according to the large manual used to classify all the neurological disorders (DSM V). In the latest edition of that manual, OCD is not classified as an anxiety disorder. Until we better understand what is going in the brain of people with OCD, I left it classified under the anxiety disorders. With OCD, a patient has obsessions which are distressing and persistent thoughts and/or compulsions which are urges to repeatedly perform a task or ritual, which are personal standards the person must adhere to in order to relieve the anxiety. These obsessions and/or compulsions are not caused by anything known and they cause stress as well as social dysfunction. The patient usually recognizes the fact that the obsessions and/or compulsions are unreasonable, but they cannot be controlled.

In developed countries, eating disorders are a major public health problem. Eating disorders can include the inability to control eating resulting in obesity or the overability to control eating with an exaggerated concern of being overweight resulting in anorexia nervosa or bulimia nervosa. Both eating disorders result in weight that is unhealthy and potentially life threatening. Our mind can override the regulatory behavior of eating based on our emotions. There is no known cause for eating disorders although it appears that genetics and environmental factors play a role. Symptoms can include fatigue, weakness, sensitivity to cold, stunted growth, acid reflux, and reduced libido.

With eating disorders, one of the key factors that can impact these eating disorders is the cultural definition and emphasis in defining what is beautiful and the cultural expectation of women and men to adhere to that definition. As a cultural, we need to abandon definitions of beauty related to physical appearance.

7.7 CONCLUSION

This chapter has discussed the basics of behavior. To better understand behavior, we looked at which animals we are closely related to through a brief review of human taxonomy. We also examined human evolution and its driving forces to better understand how human behavior evolved. We examined human emotions and moods and how they impact behavior as well as examining anxiety disorders and eating disorders.

One thing to note, no matter what is inside your skull, size doesn't matter too much at this point. Compared to *H. habilis*, size matters. But human brain size really has not changed that much in the past 1.5 million years. So compared to your sibling, size doesn't matter.

How do scientists make the claim that size doesn't matter? Some of the evidence to support that claim is in studies of Albert Einstein's brain. In 1999, the first anatomical study of his brain was published, 44 years after his death by Sandra Witelson at McMaster University in Canada, who led a team of researchers which described that Einstein's parietal lobes (implicated in mathematical cognition, as well as visual and spacial cognition) were 15% larger than the average size observed in the public. The group also reported that the overall size of Einstein's brain is on the low end of average. Later, Dean Falks research team at Florida State University examined Einstien's brain, describing other differences observed which may or may not explain the difference in intellect. Interestingly, Falk has published several papers describing human brain evolution.

Einstein's brain has reached the modern age. There is currently an iPad app called "Einstein Brain Atlas" which includes 350 images of Einstein's brain.

BIBLIOGRAPHY

Clark, D.P., Pazdernik, N.J., 2013. Molecular Biology, second ed. Elsevier Academic Press, ISBN: 978-0-12-378594-7.
Dean, F., 2009. New information about Albert Einstein's brain. Front. Evol. Neurosci. 1, 3, Published online May 4, 2009. Prepublished online April 2 2009. http://dox.doi.org/10.3389/neuro.18.003.2009 PMCID: PMC2704009.

Falk, D., Lepore, F.E., Noe, A., 2013. The cerebral cortex of Albert Einstein: a description and preliminary analysis of unpublished photographs. Brain. 136 (4), 1304–1327, Published online 2012 November 14. http://dox.doi.org/10.1093/brain/aws295 PMCID: PMC3613708.

Kandal, E.R., Schwartz, J.H., Jessel, T.M., 2000. Principles of Neural Science, fourth ed. McGraw-Hill Companies, ISBN 0-8385-7701-6.

Kolb, B., Whishaw, I.Q., 2014. An Introduction to Brain and Behavior, fourth ed. Worth Publishing, ISBN 9-781-4292-422-88.

Mason, P., 2011. Medical Neurobiology. Oxford University Press, ISBN 978-0-19-533997-0.

Men, W., Falk, D., Sun, T., Chen, W., Li, J., Yin, D., et al., 2014. The corpus callosum of Albert Einstein's brain: another clue to his high intelligence? Brain. 137 (4), e268, Published online September 21, 2013. http://dox.doi.org/10.1093/brain/awt252 PMCID: PMC3959548.

Morris, J.R., Hartl, D.L., Knoll, A.H., Lue, R.A., 2013. Biology: How Life Works. W.H. Freeman and Company, ISBN-13: 978-1-4292-1870-2.

Nicholls, J.G., Martin, A.R., Fuchs, P.A., Brown, D.A., Diamond, M.E., Weisblat, D., 2012. From Neuron to Brain, fifth ed. Sinauer Associates, Incorporated, ISBN: 978-0-87893-609-0.

Purves, D., 2012. Neuroscience, fifth ed. Sinauer Associates, Incorporated, ISBN:9780878936953.

Squire, L., Berg, D., Bloom, F., du Lac, S., Ghosh, A., Spitzer, N., 2012. Fundamental Neuroscience, fourth ed. Elsevier Academic Press, ISBN 98/0-12-385870-2.

CHAPTER 8

10:00 p.m. Counting Sheep

Contents

SUMMARY

At certain times of each day, we feel hungry or we feel sleepy. A lot of the rhythms that reoccur each day are regulated by the cycles of the day as well as the seasons. In this chapter, we are going to look at one particular rhythm, the circadian rhythm, and how that regulates our sleep—wake cycles, dreams, and what happens when normal sleeping patterns are perturbed.

8.1 BIORHYTHMS

Biorhythms are the timekeepers that our body has in place to control a multitude of biological processes that are linked to cycles of the day (day and night) and seasons. An example of one biorhythm is the human circadian rhythm, which is regulated by a part of the hypothalamus called the suprachiasmatic nucleus (SCN—each letter said individually). The SCN is found above (supra) the optic chiasm, which is how the SCN got its name. The SCN regulates circadian rhythms and body temperature in a genetic manner, meaning that the regulation of the circadian rhythm and body temperature is unlearned. The SCN comes "prewired" to regulate circadian rhythm and body temperature.

Neuroscience Basics.
DOI: http://dx.doi.org/10.1016/B978-0-12-811016-4.00008-8

8.2 CIRCADIAN RHYTHMS

Circadian rhythms are the rhythms that last a day and are synchronized to the light/dark cycle created by the Earth's rotation. Circadian rhythms—wakefulness during the day and sleep during the night, are a result of evolution. *Homo sapiens* as well as many animals evolved to maximize our ability to find our next meal. We see best during the day, so our ability to find food will be at its best during daytime. This also minimizes the energy we expend to find food, which is also advantageous.

Circadian rhythms regulate several key functions for our body. They regulate wakefulness and sleep, hormone secretion, frequency of eating and body temperature. Circadian rhythms last between 24 and 25 hours for humans. We cannot adjust to a 22- or a 28-hour day. Like many other things in our body, this rhythm is tightly regulated, including the total length of the rhythm. Circadian rhythms cannot be affected through food or water deprivation, tranquilizers, alcohol, and most brain damage. It is amazing how unshakeable circadian rhythms are.

How are circadian rhythms regulated by the brain? Well, for starters, the eye sends information to the SCN about the amount of light available in the environment. Your eye communicates the level of the outdoor light and another part of your brain interprets that information as to what time of day it is based on past experience and clocks. The SCN notifies the pineal gland about the levels of outdoor light. When the light starts to fade outside (sunset or dusk), and the lower levels of light are communicated to the pineal gland, the pineal gland secretes a chemical message—the sleep hormone (melatonin). Interestingly, melatonin secretion from the pineal gland occurs 2 hours before bedtime and ends about 2 hours before waking up. Light is not the only information that regulates circadian rhythms—noise, meals, and temperature also send information to the brain about what time of day it is.

8.2.1 Jet Lag

Jet lag is a disruption in our circadian rhythm caused by crossing time zones. It is a mismatch in the circadian rhythm and the actual external time. Jet lag can result in sleepiness during the day, sleeplessness at night, depression and impaired cognition. It can also be a major source of stress for some people.

8.3 SLEEP

Sleep is an active state defined by reduced motor activity, decreased response to stimulation and is relatively easily reversibility. Sleep is active, just as wakeful behavior is active. Some people thrash about, sleepwalk, snore, and dream, all of which are active behaviors.

Stages of sleep are defined based on the electrical activity groups (EEGs—each letter said individually) of thinking cells (neurons). There are stages of sleep that are named and classified based on the EEG generated from the brain.

Stage 1 sleep. In this stage, sleep has just started. Overall, brain activity is reduced compared to the awake state but is higher than other stages of sleep. The cortex is still receiving input and information.

Stage 2 sleep. This stage is characterized by an EEG wave known as a sleep spindle. There is a decrease in sensory input sent to the cortex. There is a decrease in heart rate and blood pressure as well as breathing rate.

Stages 3 and 4 sleep. These stages are characterized with slow wave sleep EEGs. The slow waves indicate that the neuronal activity is synchronized. By stage 4, sensory input to the cortex is massively reduced as well as heart rate, blood pressure, and breathing rate.

REM sleep. During REM, there are muscle twitches, eye movements, but other than that, muscles remain inactive. REM also has the brain activity of someone who is awake. Heart rate, breathing rate, and blood pressure are variable and may be more similar to that of an earlier stage of sleep.

Through a night, a person starts at stage 1 and progresses to stage 2 then stage 3 and then stage 4. After about an hour or so of sleep, they will cycle from stage 4 to stage 3 to stage 2 to REM. The sequence will repeat through the night, each cycle lasting about 90 minutes. Early in the night, stages 3 and 4 are dominant. Early in the morning, stage 4 grows shorter and REM is the dominant stage.

Sleep serves several functions for our body. Sleep conserves energy, reduces time spent near predators, and keeps us physically safe. And one of the key things sleep does for us is it stores our memories (Fig. 8.1). As we sleep, memories that were formed in short-term memory go to a final storage place and become a long-term memory when we get a good night's sleep. This is why as a Professor, I drive home the fact that students shouldn't pull an all-nighter studying the night before a big test. Your brain is tired, couldn't store the memories during sleep, and so you

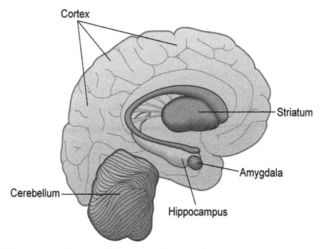

Figure 8.1 Regions of the brain necessary for learning and memory.

will actually be able to recall very little. If you want to do well on a test or a presentation, you should get a full night's sleep the night before—meaning, plan ahead as much as possible so you can sleep!

8.4 DREAMS

As we get drowsy, we fall into non-REM, tossing and turning until we are comfortable. Then eventually, we fall into REM. REM is where the dreams occur. There is not a reliable way to measure how often people dream. Most people dream every night and most people have multiple dreams per night. Dreams tend to be brief and do not last very long. Based on some initial research, it appears dreams actually occur in real time.

What is in our dreams? Both Sigmund Freud (1856–1939) and Carl Jung (1875–1961) have weighed in on their hypothesis about the nature of the content of our dreams. Freud hypothesized that the dream was symbolic events to provide a coherent account of the dreamers unconscious wishes related to sex. Carl Jung hypothesized that dreams signify distant memories encoded in the brain but lost to conscious memory, that dreams allow us to relive the history of the human race through our collective unconscious.

Why do we dream? To date, why we dream is still a mystery, the brain only knows the answer to. There are two common theories about why we dream. One theory states that dreams have no meaning and are simply a story. Another theory states that we dream as some sort of biological coping mechanism or as a way to deal with our emotions that are under the surface that we usually do not address during the day.

Do animals dream? The answer—probably. Remember we talked about the electrical recordings that characterized the stages of sleep? This has been done in several types of animals including rats and dogs. Most of us who have watched a dog sleep have assumed that they dream. They often have leg twitches like they are running or even bark in their sleep. Data from the electrical output of the brain confirms what most of us assumed—electrically, they are in a stage of sleep that would allow for dreams.

8.5 SLEEP RESEARCH IMPLICATIONS—NEUROETHICS

There are several examples of sleep research. On example demonstrates the need for ethical oversight in neuroscience research. Between 1939 and 1945, over 2 million pills of what was called pervitin and was actually crystal meth were distributed to Nazi troops to promote "wakefulness." It has also been stated that Hitler himself altered his need for sleep through the use of methamphetamines. This example demonstrates the need for neuroethics.

Neuroethics examines the implications of neuroscience on human self-understanding, ethics, law, and policy. There are neuroethicists who are examining and defining what is the responsible application of neuroscience. This has become a growing field in the past several years. This section will highlight several of the main topics being discussed among neuroethicists.

Neuroethics is growing rapidly in part because it addresses the two most important subjects relevant to who we are and how we live: the brain and the mind. There are two main types of neuroethics: primary neuroethical research and applied neuroethics. Primary neuroethics or fundamental neuroethics look at what we know about the brain (including structure and evolution) to determine how that deepens our understanding of thought, morality, and judgement. Applied neuroethics examines ethical questions that result from neuroscientific or neurotechnological advances.

Primary neuroethical research examines how our knowledge of the brain impacts our understanding of humans. An example of primary neuroethical research is cognitive enhancers. There are several medications developed for neurological disorders that improve learning and memory. Additionally, there are other interventions that can improve learning and memory. Neuroethical research is looking at several issues that arise from cognitive enhancement. The first issue is what constitutes enhancement and how is that related to disease or therapy or normal function? The second issue is do we consider cognitive enhancement as morally permissible, morally required, or morally hazardous? And one of the final issues is if cognitive enhancement is permissible, what would the social and/or global impact be of widespread cognitive enhancement use?

Primary neuroethical research also examines free will. There has been an ongoing debate among philosophers about whether or not free will is compatible with the discovery of mechanisms that regulate decision making. Some research in neuroscience suggests that our ability to make choices can be affected without our awareness. (This topic lends itself to another area of neuroethical research—neuromarketing.) Neuroethical research is examining if neuroscience can increase the ability of ethicists to attribute moral and criminal responsibility. In other words, how does our growing knowledge of the decision-making mechanisms implicate our ability to measure moral responsibility? Does neuroscience give any evidence to hold different people to different standards—juveniles or people with reduced self-control or psychopaths—who might have altered connections in decision-making areas, are they held to different standards than people who have a fully mature area for decision making?

Applied neuroethics explores different topics. Applied neuroethics examines technological advances as they occur to determine what ethical implications may arise from each of the advances—how should those advances be applied, who qualifies for those treatments, etc.? Applied neuroethics also investigates ethical implications from neuroscience research.

Annually, neuroscience receives research money from the national security organizations. How can neuroscience help? There are several brain–computer interfaces that exemplify how neuroscience can be applied to military use. There are several pieces of technology being developed that would allow monitoring the brain of fighter pilots in order to customize the cockpit to the pilots needs in real time. There are also technology that are being developed to convert subconscious neurological responses to danger into consciously available information. And what about memory enhancement? We talked briefly about this in the cognitive enhancement section above. However, what about using treatment to augment cognitive performance during a wartime situation? Or using medication developed for disorders that are characterized with sleep deprivation to keep soldiers alert during battle?

National security agencies are also using neuroscience to create advanced interrogation methods and the ability to detect deception. Increased ability to image the brain has led to enhanced methods for deception detection. This is not only being used for national security agencies, but also advertising and insurance fraud as well as the criminal justice system. Some of the methods have yet to be validated; however, these technologies raise many legal and ethical issues.

Public conversations with input from scientists about the legal and ethical implications that arise from neuroscience research must continue to grow in number and in boldness.

8.6 WHAT HAPPENS WHEN SLEEP IS DISRUPTED?

There are several disorders that are result of the brain regulating sleep: narcolepsy and insomnia. We will briefly touch on these disorders as well as mentioning some facts about shift work sleep disorder (SWSD), which is not a malfunction of the brain but a result of an altered schedule.

Insomnia is a prolonged inability to sleep. The number one characteristic of insomnia is sleepiness. Staying up late for years can impact a person's circadian rhythm and create some insomnia issues. Stress, long hours of work and irregular lifestyles and worry about sleep can impact circadian rhythms resulting in insomnia. There are other neurological disorders that may have insomnia as a side effect. Anxiety and depression account for 20–30% of the cases of insomnia reported each year. There are clinics who conduct sleep studies to determine the cause of the insomnia and tailor a treatment for each person based on the particular cause of their insomnia.

Narcolepsy is an uncontrolled ability of falling asleep at inconvenient times. Narcolepsy is characterized with cataplexy (attack of muscle weakness while the person remains awake), sleep paralysis (inability to move while falling asleep or waking up), or hallucinations (dream like experiences that are hard to distinguish from reality). These symptoms are due to the brain's inability to regulate the sleep–wake cycles. Narcolepsy can appear when a person sits still, they can automatically go into non-REM for 5–10 minutes. In people with narcolepsy, there is a reduction in the number of thinking cells (neurons) that release one of the sleep chemical messages (hypocretin/orexin) in the hypothalamus. This obviously has an impact on the ability to regulate sleep. Because the reduction in one of the sleep chemical messages (hypocretin/orexin), mutations in the gene that codes for the hypocretin/orexin result in a high probability of having narcolepsy. While these genetic mutations are not the only cause of narcolepsy, they do predict the disorder in a high percentage of cases. There are several treatment options for people with narcolepsy. There are also several documentaries of patients with narcolepsy, including *My shocking story: I woke up in a morgue* (2014).

SWSD is caused by a change in sleeping patterns. Instead of being diurnal, which most of us are, a person for reasons of work usually has to become nocturnal. SWSD primary symptom is insomnia or sleepiness due to lower quality sleep for those who work night shifts. The sleepiness can result in reduced mental cognition, irritability, and declined motor skills. The lower quality of sleep that results in a change in the secretion of the sleep hormone (melatonin). Because night shift workers are usually asleep during the day, they do not have the same exposure to the sunlight, which reduces the message from the SCN to the pineal gland, which impacts the secretion of melatonin. Because of this, people who work at night are often encouraged to take melatonin.

8.7 CONCLUSION

Biorhythms and circadian rhythms regulate daily behaviors which aids in human survival. Sleep can be measured by the activity of the thinking cells (neurons) during the different stages of sleep. There are several disorders where sleep is disrupted such as insomnia and narcolepsy. As we have mentioned in other areas of this book, sleep is necessary for the formation of new memories as well as relieving chronic stress. A good night's sleep is something we all need to function well the next day.

BIBLIOGRAPHY

Kandal, E.R., Schwartz, J.H., Jessel, T.M., 2000. Principles of Neural Science, fourth ed. McGraw-Hill Companies.
Kolb, B., Whishaw, I.Q., 2014. An Introduction to Brain and Behavior, fourth ed. Worth Publishing.

Nicholls, J.G., Martin, A.R., Fuchs, P.A., Brown, D.A., Diamond, M.E., Weisblat, D., 2012. From Neuron to Brain, fifth ed. Sinauer Associates, Incorporated.

Purves, D., 2012. Neuroscience, fifth ed. Sinauer Associates, Incorporated.

Roskies, A., 2006. Neuroscientific challenges to free will and responsibility. Trends Cognitive Science 10 (9), 419–423, Epub August 8, 2006 <http://dx.doi.org/10.1016/j.tics.2006.07.011>.

Shook, J.R., Galvagni, L., Giordano, J., 2014. Cognitive enhancement kept within contexts: neuroethics and informed public policy. Front. Syst. Neurosci. 8, 228. Available from: http://dx.doi.org/10.3389/fnsys.2014.00228.

Squire, L., Berg, D., Bloom, F., du Lac, S., Ghosh, A., Spitzer, N., 2012. Fundamental Neuroscience, fourth ed. Elsevier Academic Press, <http://www.neuroethicssociety.org>.

Tennison, M.N., Moreno, J.D., 2012. Neuroscience, ethics, and national security: the state of the art. PLoS Biol. 10 (3), e1001289. Available from: http://dx.doi.org/10.1371/journal.pbio.1001289.

INDEX

Note: Page numbers followed by "*f*" refer to figures.

Printed in the United States
By Bookmasters